D0312614

Introduction to GPS

The Global Positioning System

Second Edition

The GNSS Technology and Applications Series

Elliott Kaplan and Christopher Hegarty, Series Editors

Understanding GPS: Principles and Applications, Second Edition, Elliott Kaplan and Christopher Hegarty, editors

Introduction to GPS: The Global Positioning System, Second Edition, Ahmed El-Rabbany

GNSS Receivers for Weak Signals, Nesreen I. Ziedan

Applied Satellite Navigation Using GPS, GALILEO, and Augmentation Systems, Ramjee Prasad and Marina Ruggieri

Geographical Information Systems Demystified, Stephen R. Galati

Digital Terrain Modeling: Acquisition, Manipulation, and Applications, Naser El-Sheimy, Caterina Valeo, and Ayman Habib

For further information on these and other Artech House titles, including previously considered out-of-print books now available through our In-Print-Forever® (IPF®) program, contact:

Artech House Publishers	Artech House Books
685 Canton Street	46 Gillingham Street
Norwood, MA 02062	London SW1V 1AH UK
Phone: 781-769-9750	Phone: +44 (0)20 7596 8750
Fax: 781-769-6334	Fax: +44 (0)20 7630 0166
e-mail: artech@artechhouse.com	e-mail: artech-uk@artechhouse.com

Find us on the World Wide Web at: www.artechhouse.com

Introduction to GPS

The Global Positioning System

Second Edition

Ahmed El-Rabbany

ARTECH HOUSE

BOSTON | LONDON
artechhouse.com

Library of Congress Cataloging-in-Publication Data
A catalog record of this title is available from the Library of Congress

British Library Cataloguing in Publication Data
A catalogue record of this title is available from the British Library

Cover design by Yekatarina Ratner

© 2006 ARTECH HOUSE, INC.
685 Canton Street
Norwood, MA 02062

ISBN 10: 1-59693-016-0
ISBN 13: 978-1-59693-016-2

10 9 8 7 6

To the people who made significant contributions to my life:
my parents, my wife, and my children

Contents

	Preface	***xiii***
	Acknowledgments	*xiv*
1	**Introduction to GPS**	**1**
1.1	Overview of GPS	2
1.2	GPS Segments	3
1.3	GPS Satellite Generations	4
1.4	Current GPS Satellite Constellation	6
1.5	Control Sites	7
1.6	GPS: The Basic Idea	8
1.7	GPS Levels of Service	10
1.8	Why Use GPS?	11
	References	11
2	**GPS Details**	**13**
2.1	GPS Signal Structure	13
2.2	GPS Modernization	16

2.3   GPS Receiver Types   18

2.4   Time Systems   20

2.5   Pseudorange Measurements   21

2.6   Carrier Phase Measurements   22

2.7   Doppler Measurements   24

2.8   Cycle Slips   25

2.9   Linear Combinations of GPS Observables   26
        References   27

3   GPS Satellite Orbit   29

3.1   Motion of Space Objects   29

3.2   Types of Orbits   32

3.3   Ideal (Keplerian) Satellite Orbit   33

3.4   Perturbed Satellite Orbit   35

3.5   GPS Broadcast Orbit   36

3.6   GPS Almanac   39

3.7   Satellite Visibility   40
        References   42

4   GPS Errors and Biases   43

4.1   GPS Ephemeris Errors   44

4.2   Selective Availability   45

4.3   Satellite and Receiver Clock Errors   47

4.4   Multipath Error   48

4.5   Antenna Phase Center Variation   49

4.6   Receiver Measurement Noise   50

4.7   Ionospheric Delay   51

4.8   Tropospheric Delay   55

4.9	Other Errors and Biases	56
4.10	Satellite Geometry Measures	57
4.11	User Equivalent Range Error	60
	References	61
5	**GPS Positioning Modes**	**65**
5.1	GPS Point Positioning: The Classical Approach	66
5.2	GPS PPP	68
5.3	GPS Relative Positioning	70
5.4	Static GPS Surveying	71
5.5	Fast (Rapid) Static	73
5.6	Stop-and-Go GPS Surveying	74
5.7	RTK GPS	76
5.8	Real-Time Differential GPS	77
5.9	Real Time Versus Postprocessing	79
5.10	Communication (Radio) Link	80
	References	82
6	**Ambiguity Resolution Techniques**	**83**
6.1	Antenna Swap Method	85
6.2	OTF Ambiguity Resolution	86
	References	88
7	**GPS Data, Products, and Correction Services**	**89**
7.1	GPS Data and Product Services	91
7.2	Maritime DGPS Service	93
7.3	WADGPS Systems	95
7.4	Multisite RTK System	98
	References	99

8   GPS Standard Formats   103

8.1   RINEX Format   104

8.2   SP3 Format   109

8.3   RTCM SC-104 Standards for DGPS Services   112

8.4   NMEA 0183 Format   118
        References   121

9   GPS Integration   123

9.1   GPS/Loran-C Integration   123

9.2   GPS/LRF Integration   127

9.3   GPS/Dead Reckoning Integration   128

9.4   GPS/INS Integration   130

9.5   GPS/Pseudolite Integration   132

9.6   GPS/Cellular Integration   134
        References   136

10   GPS Applications   139

10.1   GPS for the Utility Industry   139

10.2   GPS for Forestry and Natural Resource   141

10.3   GPS for Precision Farming   142

10.4   GPS for Civil Engineering Applications   144

10.5   GPS for Monitoring Structural Deformations   145

10.6   GPS for Open-Pit Mining   146

10.7   GPS for Land Seismic Surveying   148

10.8   GPS for Marine Seismic Surveying   149

10.9   GPS for Airborne Mapping   151

10.10   GPS for Seafloor Mapping   152

10.11   GPS for Vehicle Navigation   154

10.12   GPS for Transit Systems   156

10.13   GPS for the Retail Industry   157

10.14   GPS for Cadastral Surveying   159

10.15   Waypoint Navigation (GPS Stakeout)   160
        References   161

11   Other Satellite Navigation Systems   163

11.1   The GLONASS System   163

11.2   Galileo—The European Global Satellite Navigation
        System   167

11.3   Chinese Regional Satellite Navigation System
        (Beidou System)   170

11.4   The Japanese QZSS Satellite Navigation System   172
        References   173

        Appendix A:
        Geodetic Principles—Datums, Coordinate Systems,
        and Map Projections   175

A.1   What Is a Datum?   175

A.2   Geodetic Coordinate System   177
A.2.1   Conventional Terrestrial Reference System   179
A.2.2   The WGS 84 and NAD 83 Systems   180

A.3   What Coordinates Are Obtained with GPS?   181

A.4   Datum Transformations   182

A.5   Map Projections   183
A.5.1   Transverse Mercator Projection   184
A.5.2   Universal Transverse Mercator   185
A.5.3   Modified Transverse Mercator   187
A.5.4   Lambert Conical Projection   188
A.5.5   Stereographic Double Projection   189

A.6   Local Arbitrary Mapping Systems   190

A.7   Height Systems   191
        References   193

Appendix B:
GPS Accuracy and Precision Measures   **195**
        Reference   196

Appendix C:
Useful Web Sites   **197**

C.1   GPS/Glonass/Galileo Information and Data   197

C.2   Some GPS Manufacturers   199

About the Author   **201**

Index   **203**

Preface

The idea of writing an easy-to-read, yet complete, GPS book evolved during my industrial employment term during the period from 1996 to 1997. My involvement in designing and providing short GPS courses gave me the opportunity to get direct feedbacks from GPS users with a wide variety of expertise and background. One of the most difficult tasks that I encountered was the recommendation of an appropriate GPS reference book to the course attendees. Given that the majority of the GPS users are faced with very tight schedules, it was necessary that the selected GPS book be complete and easy to read. Such a book did not exist.

Initially, I developed the vugraphs, which I used in the delivery of my short GPS courses in Canada and abroad. I then modified the vugraphs several times to accommodate not only the various types of GPS users but also my undergraduate students at both the University of New Brunswick and Ryerson University. The modified vugraphs were then used as the basis for the first edition of this GPS book. Because of the remarkable development in GPS and the development of other satellite navigation systems, both global and regional, a second revised and extended edition of the book was essential. Similar to the first edition of the book, the second edition addresses all aspects of GPS in a simple manner, avoiding any mathematics. The second edition also addresses more recent issues, such as the modernization of GPS, the global navigation satellite system (GLONASS), the Galileo satellite navigation system, the Chinese Beidou system, and the Japanese QZSS. As well, the book emphasizes GPS applications, which will benefit not only the GPS users but also the GPS marketing and sales personnel.

Chapter 1 of the book introduces the GPS system and its components. Chapter 2 examines the GPS signal structure, the GPS modernization, and the key types of the GPS measurements. The GPS satellite orbit and the broadcast ephemeris are described in Chapter 3. An in-depth discussion of the errors and biases that affect the GPS measurements, along with suggestions on how to overcome them, is presented in Chapter 4. Chapters 5 and 6 address the various modes of GPS positioning and the issue of the ambiguity resolution of the carrier-phase measurements. The various GPS services available on the market, including augmentation systems, and the standard formats used for the various types of GPS data are presented in Chapters 7 and 8, respectively. Chapter 9 focuses on the integration of the GPS with other systems, such as Loran-C, dead reckoning, inertial navigation systems, and pseudolites. The emerging micro-electro-mechanical system (MEMS)–based inertial technology is also introduced in Chapter 9. The common GPS applications in the various fields are presented in Chapter 10. The other satellite navigation systems developed or proposed in different parts of the world are covered in Chapter 11. The book ends with three appendixes. Appendix A presents some geodetic principles, including datums, coordinate systems, and map projections in a simple manner, offering a clear understanding of this widely misunderstood area. Appendix B briefly discusses the GPS accuracy and precision measures, while Appendix C lists some useful Web sites related to GPS and other satellite navigation systems.

Acknowledgments

I would like to extend my appreciation to Dr. Alfred Kleusberg, Dr. Naser El-Sheimy, Dr. David Wells, and Mr. Mahmoud Abd El-Gelil for reviewing and/or commenting on the earlier version of the manuscript. My appreciation is also extended to Artech House's anonymous peer reviewer.

1

Introduction to GPS

The first artificial satellite to be put into orbit was the Russian Sputnik 1, which was launched in 1957 to demonstrate the viability of artificial satellites. The transmitted radio signal of Sputnik 1 was monitored by researchers at the Applied Physics Laboratory (APL) of the John Hopkins University, who noticed that the satellite signal was Doppler (or frequency) shifted as result of the relative satellite motion. Following this observation, Dr. Frank T. McClure (of APL) realized that if the satellite orbit was known, a user's location could be determined based on the Doppler shift measurements. Using this concept, the APL proposed (and then developed) an innovative satellite Doppler navigation system [1, 2]. In April 1960, the first prototype satellite was successfully launched to examine the operational feasibility of the Doppler system, which was then called the Navy Navigation Satellite System (also known as Transit). Shortly after the development of the Transit system, a similar system, known as Cicada, was developed by the USSR [2]. The Transit launch program was ended in 1988, while the Transit system was retired in 1996 [1].

The Global Positioning System (GPS) is a satellite-based navigation system that was developed by the U.S. Department of Defense (DoD) in the early 1970s as the next generation replacement to the Transit system. Initially, GPS was developed as a military system to fulfill U.S. military needs. However, it was later made available to civilians, and is now a dual-use system that can be accessed by both military and civilian users [3].

GPS provides continuous positioning and timing information anywhere in the world under any weather conditions. Since GPS is a one-way-ranging (passive) system, it serves an unlimited number of users as well as being used for security reasons [4]. That is, users can only receive the satellite signals. This chapter introduces GPS, its components, and its basic idea (i.e., the operational concept).

1.1 Overview of GPS

GPS consists, nominally, of a constellation of 24 operational satellites. This constellation was completed in July 1993, which was known as the initial operational capability (IOC). The official IOC announcement, however, was made on December 8, 1993 [5]. To ensure continuous worldwide coverage, GPS satellites are arranged so that four satellites are placed in each of six orbital planes (Figure 1.1). With this constellation geometry, four to ten GPS satellites will be visible anywhere in the world, if an elevation mask of 10 degrees is considered. As discussed later, only four satellites are needed to provide the positioning, or location, information.

GPS satellite orbits are nearly circular (an elliptical shape with a maximum eccentricity of about 0.01), with an inclination of about 55 degrees to the equator. The semimajor axis of a GPS orbit is about 26,560 km (i.e., equivalent to a satellite altitude of about 20,200 km above the Earth's surface) [6]. The corresponding GPS orbital period is approximately 12 sidereal hours (~11 hours, 58 minutes). The GPS system was officially declared to have achieved full operational capability (FOC) on July 17, 1995, ensuring the availability of at least 24 operational, nonexperimental, GPS satellites. In fact, as shown in Section 1.4, since GPS achieved its FOC, the number of

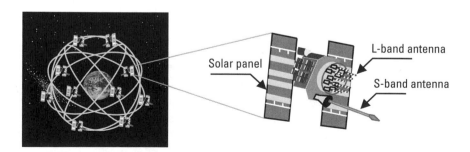

Figure 1.1 GPS constellation.

satellites in the GPS constellation has always been more than 24 operational satellites.

1.2 GPS Segments

GPS consists of three segments: space, control, and user (Figure 1.2) [7]. The space segment consists of the 24-satellite constellation introduced in the previous section. Each GPS satellite transmits a signal, which has a number of components: two sine waves (also known as carrier frequencies), two digital codes (or more for modernized GPS satellites), and a navigation message. The codes and the navigation message are added to the carriers as binary biphase modulations [7]. The carriers and the codes are used mainly to determine the distance from the user's receiver to the GPS satellites. The navigation message contains, along with other information, the coordinates (the location) of the satellites as a function of time. The transmitted signals are controlled by highly accurate atomic clocks on board the satellites. More about the GPS signal is given in Chapter 2.

The control segment of the GPS system consists of a worldwide network of tracking stations, with a master control station (MCS) located in the

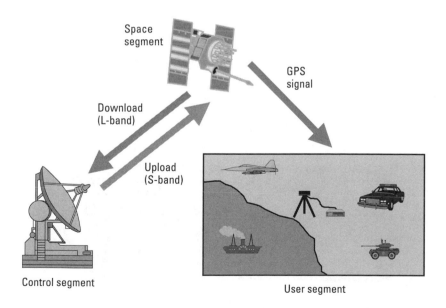

Figure 1.2 GPS segments.

United States at Schriever Air Force Base near Colorado Springs, Colorado. The primary task of the operational control segment (OCS) is to track the GPS satellites using a number of monitoring stations, in order to determine and predict satellite locations, system integrity, behavior of the satellite atomic clocks, atmospheric data, the satellite almanac, and other considerations. This information is then packed and uploaded to the GPS satellites by the ground antennas through the S-band link.

The user segment includes all military and civilian users. With a GPS receiver connected to a GPS antenna, a user can receive the GPS signals, which can be used to determine his or her position anywhere in the world. GPS is currently available to all users worldwide at no direct charge.

1.3   GPS Satellite Generations

The GPS satellite constellation buildup started with a series of 11 satellites known as Block I satellites (Figure 1.3) [8]. The first satellite in this series (and in the GPS system) was launched on February 22, 1978; the last was launched on October 9, 1985. Block I satellites were built mainly for experimental purposes. The inclination angle of the orbital planes of these satellites, with respect to the equator, was 63 degrees, which was modified in the following satellite generations. Although the design lifetime of Block I satellites was 5 years, some remained in service for more than 10 years. The last Block I satellite was taken out of service on November 18, 1995.

The second generation of the GPS satellites is known as Block II/IIA satellites (Figure 1.3). Block IIA is an advanced version of Block II, with an increase in the navigation message data storage capability from 14 days for

Block I   Block II/IIA   Block IIR

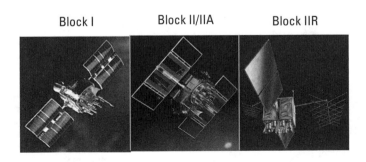

Figure 1.3 GPS   satellite   generations.   (*Source:*   www.geocities.com/mikecramer/ gps.htm/#images)

Block II to 180 days for Block IIA. This means that Block II and Block IIA satellites can function continuously, without ground support, for periods of 14 and 180 days, respectively. A total of 28 Block II/IIA satellites were launched during the period from February 1989 to November 1997. Of these, as of January 2006, 16 are in service. Unlike Block I, the orbital plane of Block II/IIA satellites is inclined at 55 degrees to the equator. The design life of a Block II/IIA satellite is 7.5 years, which was exceeded by most Block II/IIA satellites. To ensure national security, two features, known as selective availability (SA) and antispoofing (see Chapters 2 and 4), were added to Block II/IIA satellites [5, 8].

A new generation of GPS satellites, known as Block IIR and Block IIR-M ("M" stands for modernized), is currently being launched (Figure 1.3). These replenishment satellites are backward compatible with Block II/IIA, which means that any changes or differences are transparent to the users. A total of 21 Block IIR/IIR-M satellites with a design life of 7.5 years were commissioned [8]. The first Block IIR satellite was lost as a result of a launch failure on January 17, 1997. Of the remaining 20 satellites, 12 are classical Block IIR satellites and 8 are modernized Block IIR-M satellites. As of January 2006, 12 Block IIR and 1 Block IIR-M satellites have been successfully launched. In addition to the higher accuracy expected, Block IIR/IIR-M satellites have the capability of operating autonomously for at least 180 days without ground contact. The autonomous navigation capability of this generation of satellites is achieved in part through intersatellite ranging capabilities. In addition, predicted ephemeris and clock data for a period of 210 days are uploaded by the ground control segment to support autonomous navigation. An improved antenna panel was developed by Lockheed Martin for the last four Block IIRs and all of Block IIR-M satellites, which increased the power of the GPS signal received [8]. More features were added to Block IIR-M satellites under the GPS modernization program. These include the addition of civil codes (L2 CM and L2 CL codes) on the L2 frequency and two new military codes (M-codes) on both the L1 and the L2 frequencies (see Chapter 2 for details).

Block IIR/IIR-M will be followed by another constellation, called Block IIF (F stands for follow-on), consisting of 33 satellites. The satellite life span will be 12 years [8]. Block IIF satellites will have new capabilities under the GPS modernization program that will dramatically improve the autonomous GPS positioning accuracy (see Chapter 2 for details). The first Block IIF satellite is scheduled to be launched in 2007 [8]. This will be followed by GPS III with a constellation of satellites to carry GPS into 2030. GPS III, which is in the planning stage as of January 2006, is expected to provide positioning accuracy at the submeter level, real-time system integrity data, with the continuity and availability performance as required for safety-critical applications such as air navigation [8]. The first Block III satellite is scheduled to be launched in 2013.

1.4   Current GPS Satellite Constellation

The current GPS constellation (as of January 2006) contains 1 Block II, 15 Block IIA, 12 Block IIR, and 1 Block IIR-M satellites (see Table 1.1). This makes a total number of 29 GPS satellites in the constellation, which exceeds the nominal 24-satellite constellation by 5 satellites [9]. All Block I satellites are no longer operational.

The GPS satellites are placed in six orbital planes, which are labeled A through F. Since more satellites are currently available than the nominal 24-satellite constellation, an orbital plane may contain four or more satellites. The ground control has the capability to change the orbital position of any satellite. As shown in Table 1.1, the satellites can be identified by various identification systems. The most popular identification systems within the GPS user community are the space vehicle number (SVN) and the

Table 1.1
GPS Satellite Constellation as of January 2006

Sequence	SVN	PRN	Orbital Plane	Clock	Sequence	SVN	PRN	Orbital Plane	Clock
II-9	15	15	D-5	Cs	IIA-28	38	8	A-3	Cs
IIA-11	24	24	D-6	Cs	IIR-2	43	13	F-3	Rb
IIA-12	25	25	A-2	Rb	IIR-3	46	11	D-2	Rb
IIA-14	26	26	F-2	Rb	IIR-4	51	20	E-1	Rb
IIA-15	27	27	A-4	Cs	IIR-5	44	28	B-3	Rb
IIA-16	32	1	F-6	Cs	IIR-6	41	14	F-1	Rb
IIA-17	29	29	F-5	Rb	IIR-7	54	18	E-4	Rb
IIA-20	37	7	C-5	Rb	IIR-8	56	16	B-1	Rb
IIA-21	39	9	A-1	Rb	IIR-9	45	21	D-3	Rb
IIA-22	35	5	B-4	Rb	IIR-10	47	22	E-2	Rb
IIA-23	34	4	D-4	Rb	IIR-11	59	19	C-3	Rb
IIA-24	36	6	C-1	Rb	IIR-12	60	23	F-4	Rb
IIA-25	33	3	C-2	Cs	IIR-13	61	2	D-1	Rb
IIA-26	40	10	E-3	Cs	IIR-M-1	53	17	C-4	Rb
IIA-27	30	30	B-2	Rb					

Source: http://gge.unb.ca/Resources/GPSConstellationStatus.txt.

pseudorandom noise (PRN); the PRN number will be defined later. Block II/IIA satellites are equipped with four onboard atomic clocks: two cesium (Cs) and two rubidium (Rb). The cesium clock is used as the primary timing source to control the GPS signal. Block IIR satellites, however, use three rubidium clocks and no cesium clocks. Block IIF satellites are to use two rubidium and one cesium clocks [8]. It should be pointed out that two satellites, PRN05 and PRN06, are equipped with corner cube reflectors to be tracked by laser ranging (Table 1.1).

1.5 Control Sites

The control segment of GPS consists of the MCS, a worldwide network of tracking (monitor) stations, and ground antennas (Figure 1.4). The MCS, located at Schriever Air Force Base near Colorado Springs, Colorado, is the central processing facility of the control segment and is manned at all times [10].

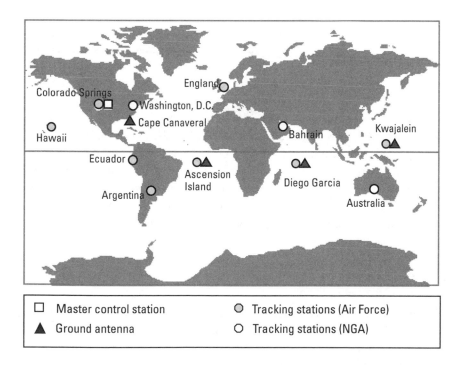

| □ | Master control station | ○ | Tracking stations (Air Force) |
| ▲ | Ground antenna | ○ | Tracking stations (NGA) |

Figure 1.4 GPS control sites.

Originally, there were five tracking stations, located in Colorado Springs (with the MCS), Hawaii, Kwajalein, Diego Garcia, and Ascension Island. However, more tracking stations from the National Geospatial-Intelligence Agency (NGA) were recently added as part of the modernization program. At the time of writing (2006), the total number of tracking stations is 12, and they are distributed in such away as to allow for the monitoring of every satellite in the GPS constellation from at least two tracking stations (Figure 1.4). The positions (or coordinates) of these monitor stations are known very precisely (i.e., presurveyed). Each monitor station is essentially equipped with high-quality GPS receivers with cesium oscillators and meteo-rological instruments for the purpose of continuous tracking of all the GPS satellites in view as well as the local meteorological conditions. Four of the tracking stations (Kwajalein, Diego Garcia, Ascension Island, and Cape Canaveral) are also equipped with ground antennas for uploading the data message (i.e., ephemeris, clock parameters, and so forth) to the GPS satellites. All of the tracking stations and the ground control stations are unmanned and operated remotely from the MCS.

The GPS observations collected at the monitor stations are transmitted via ground and satellite communication links to the MCS for processing. The outcome of the processing is predicted satellite navigation data that includes, along with other information, the satellite positions as a function of time, the satellite clock parameters, atmospheric data, satellite almanac, and others. These fresh navigation data are sent to one of the ground antennas to be uploaded to the GPS satellites through the S-band link.

Monitoring the GPS system integrity is also one of the tasks of the MCS. The status of a satellite is set to an unhealthy condition by the MCS during satellite maintenance or outages. This satellite health condition appears as a part of the satellite navigation message on a near real-time basis. Scheduled satellite maintenance or outages are reported in a message called "Notice Advisory to Navstar Users" (NANU), which is available to the public through, for example, the U.S. Coast Guard Navigation Center [9].

1.6 GPS: The Basic Idea

The idea behind GPS is rather simple. If the distances from a point on the Earth (a GPS receiver) to three GPS satellites are known along with the satellite locations, then the location of the point (or receiver) can be determined by simply applying the well-known concept of resection [11]. That is all! But how can we get the distances to the satellites as well as the satellite locations?

As mentioned before, each GPS satellite continuously transmits a microwave radio signal composed of two carriers, two codes (or more for modernized satellites), and a navigation message. When a GPS receiver is switched on, it will pick up the GPS signal through the receiver antenna. Once the receiver acquires the GPS signal, it will process it using its built-in processing software. The partial outcome of the signal processing consists of the distances to the GPS satellites through the digital codes (known as the pseudoranges) and the satellite coordinates through the navigation message.

Theoretically, only three distances to three simultaneously tracked satellites are needed. In this case, the receiver would be located at the intersection of three spheres; each has a radius of one receiver-satellite distance and is centered on that particular satellite (Figure 1.5). From a practical point of view, however, a fourth satellite is needed to account for the receiver clock offset [8]. Details on this are given in Chapter 5.

The accuracy obtained with the method described here was until May 1, 2000, limited to 100m for the horizontal component, 156m for the vertical component, and 340 ns for the time component, all at the 95 percent probability level. This low accuracy level was due to the effect of the so-called SA, a technique used to intentionally degrade the autonomous real-time positioning accuracy available to unauthorized users for security reasons [5]. However, following extensive studies, the U.S. government deactivated SA on May 2, 2000, resulting in a much-improved autonomous GPS accuracy [12, 13]. With the effect of SA reduced to zero, autonomous GPS accuracy has improved by a factor of seven or more. To further improve GPS positioning accuracy, the so-called differential or relative positioning method, which employs two receivers simultaneously tracking the same GPS satellites, is used. In this case, positioning accuracy at the level of a few meters increasing to millimeters can be obtained.

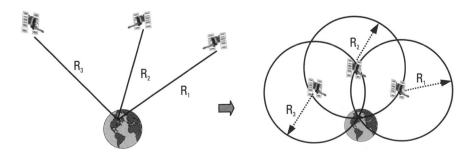

Figure 1.5 The basic idea of GPS positioning.

Other uses of GPS include the determination of the user's velocity, which could be determined by several methods. A widely used method is based on estimating the Doppler frequency of the received GPS signal (see Chapter 2 for information on Doppler measurements) [8]. It is known that the Doppler shift occurs as a result of the relative satellite-receiver motion. GPS may also be used to determine the attitude of a rigid body, such as an aircraft or a marine vessel. The word "attitude" means the orientation, or the direction, of the rigid body, which can be described by the three rotation angles of the three axes of the rigid body with respect to a reference system. Attitude is determined by equipping the body with a minimum of three GPS receivers (or one special receiver) connected to three antennas, which are arranged in a nonlinear configuration [13]. Data collected by the receivers are then processed to obtain the attitude of the rigid body.

1.7 GPS Levels of Service

As stated earlier, GPS was originally developed as a military system, but was later made available to civilians as well. However, to keep the military advantage, the U.S. DoD provides two levels of GPS positioning and timing services: the precise positioning service (PPS) and the standard positioning service (SPS) [5].

PPS is the most precise autonomous positioning and timing service. It uses one of the transmitted GPS codes, known as P(Y)-code (Chapter 2), which is accessible by authorized users only. These users include U.S. military forces and some government agencies. According to [8], the specified positioning and timing accuracies provided by the PPS is 22m for the horizontal component, 27.7m for the vertical component, and 200 ns for UTC time transfer (95 percent probability level). However, actual field measurements have shown that the positioning and timing accuracies can be as good as 8m and 10 ns, respectively [8].

SPS, on the other hand, is less precise than PPS. It uses the civil GPS code, known as the C/A-code, which is available free of charge to all users worldwide, authorized and unauthorized. Originally, SPS provided positioning accuracy of the order of 100m for the horizontal component and 156m for the vertical component (95 percent probability level). With SA deactivated, the specified positioning and timing accuracies provided by the SPS is 13m for the horizontal component, 22m for the vertical component, and 40 ns for UTC time transfer (95 percent probability level) [3]. In other words, the SPS autonomous positioning accuracy is presently at a comparable level to that of the PPS, which was confirmed through actual field measurements [8]. PPS, however, offers higher jamming resistance and signal antispoofing [14].

1.8 Why Use GPS?

GPS has revolutionized many fields (e.g., surveying and navigation) since its early stages of development. Although GPS was originally designed as a military system, its civil applications have grown much faster. As for the future, it is said that the number of GPS applications will be limited only by one's imagination.

On the surveying side, GPS has replaced conventional methods in many applications. GPS positioning has been found to be a cost-effective process, in which at least a 50 percent cost reduction can be obtained whenever it is possible to use the so-called real-time kinematic (RTK) GPS, as compared with conventional techniques [15]. In terms of productivity and time saving, GPS could provide more than 75 percent timesaving whenever it is possible to use the RTK GPS method (more about RTK capabilities and limitations is given in Chapter 5) [15]. The fact that GPS does not require intervisibility between ground stations has also made it more attractive to surveyors over the conventional methods. For those situations in which the GPS signal is obstructed, such as in urban canyons, GPS has been successfully integrated with other conventional equipment.

GPS has numerous applications in land, marine, and air navigation. Vehicle and personal tracking and navigation are rapidly growing applications. It is expected that the majority of GPS uses will be in vehicle navigation. Future uses of GPS will include automatic machine guidance and control, where hazardous areas can be mapped efficiently and safely using remotely controlled vehicles. The recent U.S. decision to modernize GPS and to deactivate the SA will undoubtedly open the door for a number of other applications yet to be developed.

References

[1]   Bowditch, N., "The American Practical Navigator," Bicentennial Edition, *NIMA*, Pub. No. 9, 2002. CD-ROM.

[2]   Parkinson, B. W., "A History of Satellite Navigation," *Navigation: Journal of The Institute of Navigation*, Vol. 42, No. 1, Special Issue, 1995, pp. 109–164.

[3]   FRP, U.S. Federal Radionavigation Plan, 2001.

[4]   Langley, R. B., "Why is the GPS Signal so Complex?" *GPS World*, Vol. 1, No. 3, May/June 1990, pp. 56–59.

[5]   Hoffmann-Wellenhof, B., H. Lichtenegger, and J. Collins, *Global Positioning System: Theory and Practice*, 5th revised edition, New York: Springer-Verlag, 2001.

[6]   Langley, R. B., "The Orbits of GPS Satellites," *GPS World*, Vol. 2, No. 3, March 1991, pp. 50–53.

[7]   Wells, D. E., et al., *Guide to GPS Positioning*, Fredericton, New Brunswick: Canadian GPS Associates, 1987.

[8]   Kaplan, E. D. and C. J. Hegarty (Eds.), *Understanding GPS: Principles and Applications*, 2nd Edition, Norwood, MA: Artech House, 2006.

[9]   U.S. Coast Guard Navigation Center, "GPS Status," January 17, 2006, http://www.navcen.uscg.gov/gps/.

[10] Leick, A., *GPS Satellite Surveying*, 2nd ed., New York: Wiley, 1995.

[11] Langley, R. B., "The Mathematics of GPS," *GPS World*, Vol. 2, No. 7, July/August 1991, pp. 45–50.

[12] Shaw, M., K. Sandhoo, and D. Turner, "Modernization of the Global Positioning System," *GPS World*, Vol. 11, No. 9, September 2000, pp. 36–44.

[13] Conley, R., "Life After Selective Availability," *U.S. Institute of Navigation Newsletter*, Vol. 10, No. 1, Spring 2000, pp. 3–4.

[13] Kleusberg, A., "Mathematics of Attitude Determination with GPS," *GPS World*, Vol. 6, No. 9, September 1995, pp. 72–78.

[14] Misra, P. and P. Enge, *Global Positioning System—Signals, Measurements, and Performance*, Lincoln, MA: Ganga-Jamuna Press, 2001.

[15] Berg, R. E., "Evaluation of Real-Time Kinematic GPS Versus Total Stations for Highway Engineering Surveys," 8th Intl. Conf. Geomatics: Geomatics in the Era of RADARSAT, Ottawa, Canada, May 24–30, 1996, CD-ROM.

2

GPS Details

Positioning, or finding the user's location, with GPS requires some understanding of the GPS signal structure and how the measurements can be
made. Likewise, as the GPS signal is received through a GPS receiver, understanding the capabilities and limitations of the various types of GPS receivers
is essential. Furthermore, the GPS measurements, like all measurable quanti-
ties, contain errors and biases, which can be removed or reduced by combining the various GPS observables. This chapter discusses these issues in detail.

2.1   GPS Signal Structure

When describing the GPS signal structure, we must distinguish between traditional GPS satellites (Blocks II, IIA, and IIR) and modernized GPS satellites (Blocks IIR-M and subsequent blocks of GPS satellites). As mentioned
in Chapter 1, each GPS satellite transmits a microwave radio signal composed of two carrier frequencies (or sine waves) modulated by two digital
(ranging) codes (or more for modernized GPS satellites) and a navigation
message (see Figure 2.1). The two carrier frequencies are generated at
1,575.42 MHz (referred to as the L1 carrier) and 1,227.60 MHz (referred to
as the L2 carrier). The corresponding carrier wavelengths are approximately
19 cm and 24.4 cm, respectively, which result from the relation between the
carrier frequency and the speed of light in space [1, 2]. The availability of the
two carrier frequencies allows for correcting a major GPS error, known as the

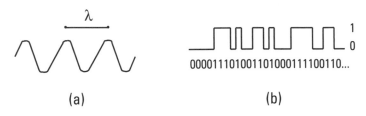

Figure 2.1   (a) A sinusoidal wave; (b) a digital code.

ionospheric delay (see Chapter 4 for details). All of the GPS satellites transmit the same L1 and L2 carrier frequencies. The code modulation, however, is different for each satellite, which significantly minimizes the signal interference.

Traditional GPS satellites transmit two ranging codes known as coarse acquisition (or C/A-code) and precision (or P-code). The C/A-code is modulated onto the L1 carrier only, while the P-code is modulated onto both the L1 and the L2 carriers. Modernized GPS satellites, however, transmit two additional new ranging codes, namely the L2 civil-moderate (L2 CM) code and L2 civil-long (L2 CL) code [3]. Each of the ranging codes consists of a stream of binary digits, zeros, and ones, known as bits or chips. The codes are commonly known as PRN codes because they look like random signals (i.e., they are noise-like signals). But in reality, the codes are generated using a mathematical algorithm. GPS signal modulation is called biphase modulation, because the carrier phase is shifted by 180 degrees when the code value changes from zero to one or from one to zero [4].

The C/A-code is a stream of 1,023 binary digits (i.e., 1,023 zeros and ones) that repeats itself every millisecond (ms). This means that the chipping rate of the C/A-code is 1.023 Mbps. In other words, the duration of one bit is approximately 1 μs, or equivalently 300m [5]. Each satellite is assigned a unique C/A-code, which enables the GPS receivers to identify which satellite is transmitting a particular code. Presently, 37 C/A-codes are defined; of which 36 are mutually exclusive (as C/A codes 34 and 37 are common) [3]. The C/A-code range measurement is relatively less precise compared with that of the P-code. It is, however, less complex and is available to all users.

The P-code is a very long sequence of binary digits that repeats itself after 266 days [1]. It is also 10 times faster than the C/A-code (i.e., its rate is 10.23 Mbps). Multiplying the time it takes the P-code to repeat itself, 266 days, by its rate, 10.23 Mbps, tells us that the P-code is a stream of about 2.35×10^{14} chips! The 266-day-long code is divided into 38 segments; each is one week long. Of these, 32 segments are assigned to the various GPS satellites. That is, each satellite transmits a unique one-week segment of the P-code, which is initialized every Saturday/Sunday midnight crossing. In fact, each satellite is assigned a particular pair of C/A-code and P-code, which are inseparable [3]. The remaining six segments of the P-code are reserved for other uses (e.g., ground transmitters). It is worth mentioning that a GPS satellite is usually identified by its unique one-week segment of the P-code. For example, a GPS satellite with an ID of PRN 20 refers to a GPS satellite that is assigned the twentieth-week segment of the PRN P-code. The P-code is designed primarily for military purposes. It was available to all users until January 31, 1994 [1]. At that time, the P-code was encrypted by adding to it an unknown W-code. The resulting encrypted code is called the Y-code, which has the same chipping rate as the P-code. This encryption is known as antispoofing (AS).

As indicated earlier, modernized GPS satellites transmit two new PRN codes known as the L2 CM and L2 CL codes. Both codes are transmitted at a chipping rate of 511.5 Kbps [3]. The L2 CM code, however, is a stream of 10,230 chips with a length of 20 ms, while the L2 CL code is a stream of 767,250 chips with a length of 1.5s. In other words, the L2 CL code is 75 times longer than L2 CM code. The overall chipping rate of the L2 civil (called L2C) signal is twice the chipping rate of the CM and CL (i.e., 1.023 Mbps, which is the same as the L1 C/A-code).

The GPS navigation message is a low rate of 50 bps data stream, which after being binary (modulo-2) added to the P(Y)- and C/A-codes, are modulated onto both the L1 and the L2 carriers. A complete navigation message consists of 25 frames of 1,500 bits each, or 37,500 bits in total (i.e., a complete navigation message takes 750 seconds, or 12.5 minutes, to be transmitted). Each of the frames is divided into 5 subframes of 300 bits each. The navigation message contains, along with other information, the coordinates of the GPS satellites as a function of time (broadcast ephemeris), the satellite clock correction model parameters, the satellite health status, the satellite almanac, and atmospheric (ionospheric) correction model parameters. To help the receiver identify which part of the long P- (or Y-) code is being transmitted, and consequently lock rapidly to it, the navigation message contains an important part known as the handover word (HOW). Typically, a GPS receiver locks on the C/A-code first and then retrieves the navigation message, which contains the HOW in each subframe. Each satellite transmits its own navigation message with information on the other satellites, such as the approximate location and health status [1]. The navigation message is common to the P(Y)- and C/A-codes on both the L1 and L2 channels [3].

For Block IIR-M, the navigation message is binary (modulo-2) added to the L2 CM code only. One of two different rates, 50 bps or 50 symbols per second (sps), can be used, depending on ground command selection. The resultant is then combined with the L2 CL code using the so-called time-division multiplexing technique and then modulated onto the L2 carrier. Another possibility of modulation mode, which is again selectable by ground command, is to modulate the L2 carrier by the binary (modulo-2) addition of the C/A-code and the navigation message [3].

2.2 GPS Modernization

The current GPS signal structure was designed in the early 1970s, more than 30 years ago [6]. In the next 20 years, the GPS constellation is expected to have a combination of Block IIR/IIR-M satellites, Block IIF, and possibly Block III satellites. To meet the future requirements, the GPS decision makers have studied several options to adequately modify the signal structure and system architecture of the future GPS constellation. The modernization program aims to, among other things, provide signal redundancy and improve positioning accuracy, signal availability, and system integrity.

One component of the modernization program is related to upgrading the ground control facilities of the GPS system. In 1987, Wells et al. reported that the accuracy of the broadcast orbit was typically 20m and could occasionally reach 80m [4]. In 2005, however, the orbital accuracy was reported to have improved to about 2m (i.e., an improvement by one order of magnitude). Such an accuracy improvement resulted from two phases of a five-phase modernization process, namely (1) improvement in the operational software, and (2) better orbital modeling. The third phase, introduced recently (September 2005), included the addition of six new monitoring stations to improve the GPS broadcast ephemeris and clock data. As a result, it is expected that real-time navigation accuracy will improve by 15 to 20 percent [7]. With the completion of the third phase, satellite operators will be able to monitor every satellite in the GPS constellation from at least two monitoring stations. Five more stations will be added in the future, which will allow the satellite operators to monitor every satellite in the GPS constellation from at least three monitoring stations. This will result in great improvement to the satellite integrity monitoring. Phases four and five will test the backup facility of the MCS and follow-up on the modeling upgrade, respectively [7].

The modernization program also includes the addition of civil codes (L2 CM and L2 CL codes) on the L2 frequency and two new military codes (M-codes) on both the L1 and the L2 frequencies [6]. As such, the modernized L2 is shared between military and civil (L2C) signals. The new codes are added to a modernized generation of Block IIR satellites, known as IIR-M, as explained in Section 2.1. The first modernized GPS satellite (IIR-M1) was successfully launched on September 25, 2005. The additional signals transmitted by modernized IIR-M GPS satellites offer a number of advantages, including improved interference resistance, enhanced tracking, and better positioning performance indoors and in forest areas. Most importantly, the addition of a civil code on the L2 frequency makes it possible to recover the full L2 carrier without the need to use current inefficient tracking techniques, such as cross-correlation (see Section 2.3). In fact, the availability of civil codes on both L1 and L2 frequencies allows a user with a stand-alone GPS receiver to correct for the effect of the ionosphere (the upper layer of the atmosphere), which is a major error source (see Chapter 4 for details). With the deactivation of SA and the upgrade of the GPS tracking system, it is expected that once a sufficient number of satellites with the new capabilities is available, the autonomous GPS accuracy will be at the few meter level or better.

Although it improves the autonomous GPS accuracy, the addition of a second civil signal is found to be insufficient for use in the civil aviation safety-of-life applications. This is mainly because of the potential interference from the ground radars that operate near the GPS L2 band. As such, to satisfy aviation user requirements, a third civil signal at 1,176.45 MHz (called L5) will be added, along with the L2C and the M-code on L1 and L2, to future Block IIF satellites as part of the modernization program [6]. The third frequency will be robust and will have a higher power level. Two PRN ranging codes will be transmitted on L5, which are known as I5-code and Q5-code, respectively. Both codes are transmitted at a chipping rate of 10.23 Mbps with a length of 1 ms. The navigation data is a 100-sps symbol stream, which is modulo-2 added to the I5-code only. No navigation data will be added to Q5-code [8].

The modernization of GPS will also include the studies for the next generation Block III satellites, which will carry GPS into 2030. The addition of these capabilities will dramatically improve the autonomous GPS positioning accuracy. As well, RTK users, who require centimeter-level accuracy in real time, will be able to resolve the initial integer ambiguity parameters instantaneously. More about RTK positioning is given in Chapter 5.

2.3   GPS Receiver Types

In 1980, only one commercial GPS receiver was available on the market, at a price of several hundred thousand U.S. dollars [6]. This, however, has changed considerably, as more than 500 different GPS receivers from more than 70 companies are available in today's market (see, for example, the January 2005 issue of *GPS World* magazine). The current receiver price varies from about $100 for the simple handheld units to about $15,000 for the sophisticated geodetic-grade units. The price will continue to decline in the future as the receiver technology becomes more advanced, among other factors. A GPS receiver requires an antenna to be attached to it, either internally or externally. The antenna receives the incoming satellite signal and then converts its energy into an electric current, which can be handled by the GPS receiver [9, 10].

First generation GPS receivers utilized analog circuitry, which made them bulky and heavy. Recently, however, as a result of evolution in semiconductor technology and the use of digital circuitry, GPS receivers are now considerably smaller and lighter. A new generation of GPS receivers, called software GPS (or GNSS) receivers, is now on the horizon. Software receivers use reprogrammable software, rather than hardware, for signal acquisition and processing [11]. As such, they provide more flexibility and cost effectiveness, particularly now that new signals from modernized GPS satellites and other satellite-based navigation systems are becoming available.

Commercial GPS receivers may be divided into four types, according to their receiving capabilities: single-frequency code receivers, single-frequency carrier-smoothed code receivers, single-frequency code and carrier receivers, and dual-frequency receivers. Single-frequency receivers access the L1 frequency only, while dual-frequency receivers access both the L1 and the L2 frequencies. Figure 2.2 shows examples of various types of GPS receivers. GPS receivers can also be categorized according to their number of tracking channels. A good GPS receiver would be multichannel, with each channel dedicated to continuously tracking a particular satellite. Presently, most GPS receivers have 9 to 12 independent (or parallel) channels. Some receivers, which track more than one satellite system, have even more channels. For example, Javad Navigation introduced a new chip receiver, GeNiuSS, which has 72 channels and can track the GPS constellation, GLONASS (the Russian GPS counterpart), the European Galileo system now under development, and two augmentation systems known as WAAS and EGNOS (see Chapters 7 and 11 for details). Special features such as raw data output rate and solution latency are important for some applications (e.g., hydrographic

Magellan handheld
GPS receiver

Ashtech ZX geodetic quality
GPS receiver

Figure 2.2 Examples of various types of GPS Receivers. *Courtesy of:* Thales Navigation.

surveying). Other features, such as cost, ease of use, power consumption, size and weight, internal and/or external data-storage capabilities, communication options (e.g., access to the Internet and wireless receiver communication and control), interfacing capabilities, and multipath mitigation, are to be considered when selecting a GPS receiver.

The first receiver type, the single-frequency code receiver, measures the pseudoranges with the C/A-code only. No other measurements are available. It is the least expensive and the least accurate receiver type, and is mostly used for recreation purposes. The second receiver type, the single-frequency carrier-smoothed code receiver, also measures the pseudoranges with the C/A-code only. However, with this receiver type, the higher-resolution carrier frequency is used internally to improve the resolution of the code pseudorange, which results in high-precision pseudorange measurements. Single-frequency code and carrier receivers output the raw C/A-code pseudoranges, the L1 carrier-phase measurements, and the navigation message. In addition, this receiver type is capable of performing the functions of the other receiver types discussed earlier.

Dual-frequency receivers are the most sophisticated and most expensive receiver type. Before the activation of AS, dual-frequency receivers were capable of outputting all of the GPS signal components (i.e., L1 and L2 carriers, C/A-code, P-code on both L1 and L2, and the navigation message). However, after the activation of AS, the P-code was encrypted to Y-code. This means that the receiver cannot output either the P-code or the L2 carrier using the traditional signal-recovering technique. To overcome this problem,

GPS receiver manufacturers invented a number of techniques that do not require information of the Y-code. At the present time, most receivers use two techniques known as the Z-tracking and the cross-correlation techniques. Both techniques recover the full L2 carrier, but at a degraded signal strength. The amount of signal strength degradation is higher in the cross-correlation techniques compared with the Z-tracking technique. As indicated earlier, tracking the modernized GPS satellites (IIR-M and subsequent blocks) does not require the use of these tracking techniques. It is interesting to know that a number of GPS receivers with L2C capability are already available on the market. An example of these is the Trimble R7 dual-frequency GPS receiver.

2.4 Time Systems

Time plays a very important role in positioning with GPS. As explained in Chapter 1, the GPS signal is controlled by accurate timing devices, the atomic satellite clocks [12]. In addition, measuring the ranges (distances) from the receiver to the satellites is based on both the receiver and the satellite clocks. GPS is also a timing system (i.e., it can be used for time synchronization).

A number of time systems are used worldwide for various purposes [1]. Of these, the Coordinated Universal Time (UTC) and the GPS Time are the most important to GPS users. UTC is an atomic time scale based on the International Atomic Time (TAI). TAI is a uniform time scale, which is computed based on independent time scales generated by atomic clocks located at various timing laboratories throughout the world. In surveying and navigation, however, a time system with relation to the rotation of the Earth, not the atomic time, is desired. This is achieved by occasionally adjusting the UTC time scale by 1s increments, known as leap seconds, to keep it within 0.9s of another time scale called the Universal Time 1 (UT1) [12, 13], where UT1 is a universal time that gives a measure of the rotation of the Earth. Leap seconds are introduced occasionally, on either June 30 or December 31. As of May 2006, the last leap second was introduced on January 1, 2006, which made the difference between TAI and UTC time scales exactly 33s (TAI is ahead of UTC). However, discussions on the redefinition of UTC, including a possible suppression of the leap seconds procedure, are currently underway. More information about the leap seconds can be found at the U.S. Naval Observatory Web site, (http://maia.usno.navy.mil).

GPS Time is the time scale used for referencing, or time tagging, the GPS signals. It is computed based on the time scales generated by the atomic clocks at the monitor stations and onboard GPS satellites. There are no leap seconds introduced into GPS Time, which means that GPS Time is a continuous time scale. GPS Time scale was set equal to that of the UTC on January 6, 1980 [12]. However, due to the leap seconds introduced into the UTC time scale, GPS Time moved ahead of the UTC by 14 seconds on January 1, 2006. Figure 2.3 shows a history of the difference between GPS Time and UTC. The latest difference between GPS and UTC time scales is given in the GPS navigation message. It is worth mentioning that, as shown in Chapter 4, both GPS satellite and receiver clocks are offset from the GPS Time, as a result of satellite and receiver clock errors. It is also worth defining the GPS week number, which represents the number of weeks elapsed since the week of January 6, 1980, which is counted as week #0.

2.5 Pseudorange Measurements

The pseudorange is a measure of the range, or distance, between the GPS receiver and the GPS satellite (more precisely, it is the distance between the antenna center of the GPS receiver and the antenna center of the GPS satellite). As stated before, the ranges from the receiver to the satellites are needed

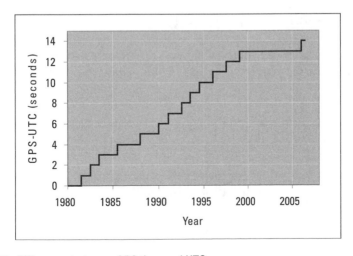

Figure 2.3 Difference between GPS time and UTC.

for the position computation. Either the P(Y)-code or the C/A-code can be used for measuring the pseudorange.

The procedure of the GPS range determination, or pseudoranging, can be described as follows. Let us assume for a moment that both the satellite and the receiver clocks, which control the signal generation, are perfectly synchronized with each other. When the PRN code is transmitted from the satellite, the receiver generates an exact replica of that code [4]. After some time, equivalent to the signal travel time in space, the transmitted code will be picked up by the receiver. By comparing the transmitted code and its replica, the receiver can compute the signal travel time. Multiplying the travel time by the speed of light (299,729,458 m/s) gives the range between the satellite and the receiver. Figure 2.4 explains the pseudorange measurements.

Unfortunately, the assumption that the receiver and satellite clocks are synchronized is not exactly true. In fact, the measured range is contaminated by the synchronization error between the satellite and receiver clocks, along with other errors and biases. For this reason, this quantity is referred to as the *pseudorange*, rather than the range [5].

GPS was designed so that the range determined by the civilian C/A-code would be less precise than that of military P-code. This is based on the fact that the resolution of the C/A-code, 300m, is 10 times lower than the P-code. Surprisingly, due to the improvements in the receiver technology, the obtained accuracy was almost the same from both codes [5].

2.6 Carrier Phase Measurements

Another way of measuring the ranges to the satellites can be obtained through the carrier phases. The range would simply be the sum of the total

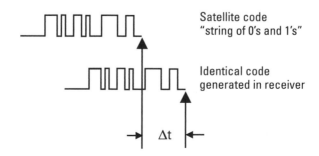

Figure 2.4 The pseudorange measurements. (*Note:* Δt = signal travel time.)

number of full carrier cycles plus fractional cycles at the receiver and the satellite, multiplied by the carrier wavelength (see Figure 2.5). The ranges determined with the carriers are far more accurate than those obtained with the codes (i.e., the pseudoranges) [5]. This is due to the fact that the wavelength (or resolution) of the carrier phase, 19 cm in the case of L1 frequency, is much smaller than those of the codes.

There is, however, one problem. The carriers are just pure sinusoidal waves, which mean that all cycles look the same. Therefore, a GPS receiver has no means to differentiate one cycle from another [5]. In other words, the receiver, when it is switched on, cannot determine the total number of the complete cycles between the satellite and the receiver. It can only measure a fraction of a cycle very accurately (less than 2 mm), while the initial number of complete cycles remains unknown, or ambiguous. This is, therefore, commonly known as the initial integer cycle ambiguity, or the ambiguity bias. Fortunately, the receiver has the capability to keep track of the phase changes after being switched on. This means that the initial cycle ambiguity remains unchanged over time, as long as no signal loss (or cycle slips) occurs [4].

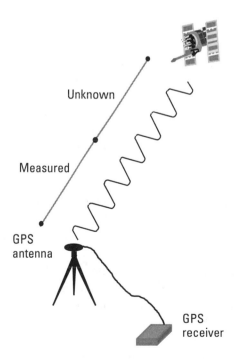

Figure 2.5 The carrier phase measurements.

It is clear that if the initial cycle ambiguity parameters are resolved, accurate range measurements can be obtained, which lead to accurate position determination. This high-accuracy positioning can be routinely achieved through the so-called relative (or differential) positioning techniques, either in real time or in the postprocessing mode. Unfortunately, this requires two GPS receivers simultaneously tracking the same satellites in view. More recently, the concept of precise point positioning (PPP) was introduced, which results in high-accuracy positioning (decimeter to centimeter level) for stand-alone GPS receivers. More about the various positioning techniques and the ways of resolving the ambiguity parameters is given in Chapters 5 and 6, respectively.

2.7 Doppler Measurements

A common phenomenon with which all of us are familiar is that we notice a change in pitch of a train whistle as the train moves past us. The change in pitch increases as the train approaches us and decreases as the train moves away from us. Well, such a change in pitch is in fact due to a physical phenomenon known as the Doppler effect, or frequency shift, named after the Austrian physicist Christian Doppler (1803–1853) [14]. In this example, the Doppler effect represents the difference in frequency of the acoustic signal received at the observer (i.e., us) and the frequency at the source (i.e., the train). In general, Doppler effect (or frequency shift) occurs as a result of a relative motion (or velocity) between a source of sound (or radiation) and an observer (or receiver). Doppler effect is used in many applications, including marine navigation, astronomy, and even detection of speedy cars using so-called *radar guns.*

As a result of the relative motion between the GPS satellites and a GPS receiver (whether stationary or in low-dynamic motion), the received GPS signal will be Doppler (or frequency) shifted. As such, a relationship can be established that relates, among other known parameters, the amount of Doppler shift, the satellite velocity, and the receiver's velocity [15]. Some GPS receivers provide Doppler measurements of the received GPS signals, which can be used, along with an estimate of satellite velocity through ephemeris information, to determine the receiver velocity. However, it should be pointed out that, because of the high altitude of GPS satellites, the satellite-receiver relative motion (and consequently the Doppler shift) will not be significant. As such, the Doppler-based receiver's velocity may not be accurate enough for some applications.

2.8 Cycle Slips

A cycle slip is defined as a discontinuity or a jump in the GPS carrier phase measurements, by an integer number of cycles, caused by temporary signal loss [1]. Signal loss is caused by obstruction of the GPS satellite signal due to buildings, bridges, trees, and other objects (Figure 2.6). This is mainly because the GPS signal is a weak and noisy signal. Radio interference, severe ionospheric disturbance, and high receiver dynamics can also cause signal loss. Cycle slips could occur due to a receiver malfunction [1].

Cycle slips may occur briefly or may remain for several minutes or even more. Cycle slips could affect one or more satellite signals. The size of a cycle slip could be as small as one cycle or as large as millions of cycles. Cycle slips must be identified and corrected to avoid large errors in the computed coordinates. This can be done using several methods—for example, by examining the so-called triple difference observable (see Section 2.9). A cycle slip will only affect one *triple difference* and therefore will appear as a spike in the triple difference data series. In some extreme cases, such as severe ionospheric activities, it might be difficult to correctly detect and repair cycle slips using triple difference observable [1, 4]. Visual inspection of the adjustment residuals might be useful to locate any remaining cycle slip.

Figure 2.6 GPS cycle slips.

As shown in Chapter 4, a zero baseline test is used to detect cycle slips due to receiver malfunction. In this test, two receivers are connected to one antenna through a signal splitter. Cycle slips can be detected by examining the adjustment residuals [4].

2.9   Linear Combinations of GPS Observables

GPS measurements are corrupted by a number of errors and biases (discussed in detail in Chapter 4), which are difficult to model fully. The unmodeled errors and biases limit the positioning accuracy of the stand-alone GPS receiver. Fortunately, GPS receivers in close proximity will share to a high degree of similarity the same errors and biases. As such, for those receivers, a major part of the GPS error budget can simply be removed by combining their GPS observables.

In principle, there are three groups of GPS errors and biases: satellite-related, receiver-related, and atmospheric errors and biases [4]. The measurements of two GPS receivers simultaneously tracking a particular satellite contain more or less the same satellite-related errors and atmospheric errors. The shorter the separation between the two receivers is, the more similar the errors and biases. Therefore, if we take the difference between the measurements collected at these two GPS receivers, the satellite-related errors and the atmospheric errors will be reduced significantly. In fact, as shown in Chapter 4, the satellite clock error is effectively removed with this linear combination. This linear combination is known as between-receiver single difference (Figure 2.7).

Similarly, the two measurements of a single receiver tracking two satellites contain the same receiver clock errors. Therefore, taking the difference between these two measurements removes the receiver clock errors. This difference is known as between-satellite single difference (Figure 2.7).

When two receivers track two satellites simultaneously, two between-receiver single difference observables could be formed. Subtracting these two single difference observables from each other generates the so-called double difference [4]. This linear combination removes the satellite and receiver clock errors. The other errors are greatly reduced. In addition, this observable preserves the integer nature of the ambiguity parameters. It is therefore used for precise carrier phase–based GPS positioning.

Another important linear combination in known as the *triple difference*, which results from differencing two double-difference observables over two epochs of time [4]. As explained in the previous section, the ambiguity

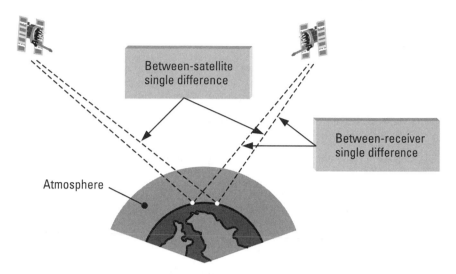

Figure 2.7 Some GPS linear combinations.

parameters remain constant over time, as long as there are no cycle slips. As such, when forming the triple difference, the constant ambiguity parameters disappear. If, however, there is a cycle slip in the data, it will affect one triple-difference observable only, and therefore will appear as a spike in the triple-difference data series. It is for this reason that the triple-difference linear combination is used for detecting the cycle slips.

All of these linear combinations can be formed with single-frequency data, whether it is the carrier phase or the pseudorange observables. If dual-frequency data is available, other useful linear combinations could be formed. One such linear combination is known as the ionosphere-free linear combination. As shown in Chapter 4, ionospheric delay is inversely proportional to the square of the carrier frequency. Based on this characteristic, the ionosphere-free observable combines the L1 and L2 measurements to essentially eliminate the ionospheric effect. The L1 and L2 carrier-phase measurements could also be combined to form the so-called wide-lane observable, an artificial signal with an effective wavelength of about 86 cm. This long wavelength helps in resolving the integer ambiguity parameters [1].

References

[1] Hoffmann-Wellenhof, B., H. Lichtenegger, and J. Collins, *Global Positioning System: Theory and Practice,* 5th revised edition, New York: Springer-Verlag, 2001.

[2] Langley, R. B., "Why Is the GPS Signal So Complex?" *GPS World*, Vol. 1, No. 3, May/June 1990, pp. 56–59.

[3] ICD-GPS-200C, "Interface Control Document ICD-GPS-200, Revision C," IRN-200C-005R1, ARINC Research Corporation, El Segundo, CA, 14 January 2003, http://www.navcen.uscg.gov/gps/modernization/default.htm.

[4] Wells, D. E., et al., *Guide to GPS Positioning*, Fredericton, New Brunswick: Canadian GPS Associates, 1987.

[5] Langley, R. B., "The GPS Observables," *GPS World*, Vol. 4, No. 4, April 1993, pp. 52–59.

[6] Shaw, M., K. Sandhoo, and D. Turner, "Modernization of the Global Positioning System," *GPS World*, Vol. 11, No. 9, September 2000, pp. 36–44.

[7] Global View, "NGA Data Drives Improved Accuracy," *GPS World*, Vol. 16, No. 10, October 2005, p. 14.

[8] ICD-GPS-705, "Interface Control Document ICD-GPS-705, Revision 2," ARINC Research Corporation, El Segundo, CA, 2 December 2002, http://www.navcen.uscg.gov/gps/modernization/default.htm.

[9] Langley, R. B., "The GPS Receiver: An Introduction," *GPS World*, Vol. 2, No. 1, January 1991, pp. 50–53.

[10] Langley, R. B., "Smaller and Smaller: The Evolution of the GPS Receiver," *GPS World*, Vol. 11, No. 4, April 2000, pp. 54–58.

[11] MacGougan, G., P. L. Normark, and C. Stahlberg, "Satellite Navigation Evolution—The Software GNSS Receiver," *GPS World*, Vol. 16, No. 10, October 2005, pp. 48–55.

[12] Langley, R. B., "Time, Clocks, and GPS," *GPS World*, Vol. 2, No. 10, November/December 1991, pp. 38–42.

[13] McCarthy, D. D., and W. J. Klepczynski, "GPS and Leap Seconds: Time to Change," *GPS World*, Vol. 10, No. 11, November 1999, pp. 50–57.

[14] Seeber, G., *Satellite Geodesy*, 2nd revised edition, Berlin, Germany: Walter de Gruyter, 2003.

[15] Kaplan, E., (ed.), Understanding GPS Principles and Applications, Norwood, MA: Artech House, 1996.

3

GPS Satellite Orbit

The ability to determine a user's position at any time requires accurate information about the positions of GPS satellites at the moments of signal transmission. As shown in Chapter 2, the coordinates of the GPS satellites as a function of time (or ephemeris) are transmitted as part of the navigation message. In fact, satellite coordinates are represented in the broadcast ephemeris in the form of predicted orbital parameters, which can be used to estimate future positions of GPS satellites. The ability to accurately predict the satellite orbit requires comprehensive knowledge of orbital mechanics, which is presented in a simple form in this chapter. First, the motion of space objects is briefly described, followed by a brief description of the common satellite orbits. The ideal satellite orbit (so-called Keplerian orbit) is described in Section 3.3. Actual (or perturbed) satellite orbits and their parameters as included in the broadcast ephemeris are then presented. A description of the GPS almanac and its use in mission planning is presented in Section 3.6. This chapter ends with a discussion on satellite visibility and the determination of visible number of satellites using the satellite almanac.

3.1 Motion of Space Objects

The amazing night sky has inspired humankind since the beginning of creation. The nighttime sky appeared to be changing; planets appeared to be in motion and to vary in brightness. Astronomers were faced with the

challenges of not only describing the motions of the planets but also determining whether such motions are related to the brightness of the planets [1].
Among these was a Danish astronomer with the name Tycho Brahe
(1546–1601), who used instruments of his own design (before the invention
of the telescope) to follow the paths of planets, particularly Mars. After many
years of accumulating quality astronomical observations, Brahe was joined by
Johannes Kepler (1571–1630), a German mathematician and astronomer, in
a trial to find a theory that explains the behavior of the planetary data. A year
later Brahe died, leaving invaluable accumulated astronomical observations
in Kepler's possession.

With the acceptance of the solar system as heliocentric (i.e., the Sun is
located at its center) and that the planets orbit the Sun in circular orbits, as
proposed by Copernicus (1473–1543), Kepler started to make use of Brahe's
planetary data [1, 2]. After trying for a long time, Kepler discovered that the
orbital paths of the planets are not circular. Instead, he found that the planets
orbit the Sun in elliptical (i.e., flattened circular) orbits, with the Sun located
at one of the ellipse's focal points (see Figure 3.1 for the Earth's orbit around
the Sun). This finding became what is known as *Kepler's first law of planetary
motion*, which states that the orbital path of a planet takes the shape of an
ellipse, with the Sun located at one of its focal points [1, 2]. Within the solar
system, with the exceptions of Mercury and Pluto's paths, planets orbit the
Sun in slightly elliptical (i.e., near-circular) orbits.

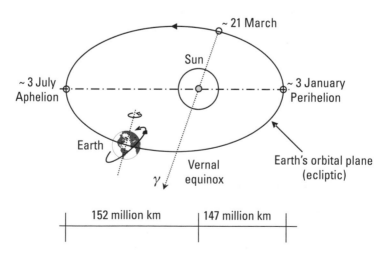

Figure 3.1 Earth's orbital plane around the Sun. (*Note:* vernal equinox represents the
imaginary line of intersection between the Earth's orbital plane and the
Equator).

Kepler continued with his work by plotting the positions of a particular planet on its orbit at different times. Surprisingly, he found that the (imaginary) line connecting the Sun to any planet sweeps out equal areas of the orbital ellipse in equal time intervals (see Figure 3.2). This became *Kepler's second law of planetary motion*. Since the wedge-shaped areas are equal and the line connecting the Sun to a planet changes, then the speed of the planet changes over time. As shown in Figure 3.1, the Earth is closer to the Sun in the winter than it is in summer, which means that the Earth moves faster around the Sun in winter.

Kepler tried to find a relationship between a planet's orbital period (i.e., the time it takes the planet to complete one revolution around the Sun) and its mean distance from the Sun. Neither a linear nor quadratic relationship gave the right fit to the collected planetary observations. A good fit, however, was found when he considered that the ratio of the square of the planet's orbital period and the cube of the mean distance from the Sun is constant. This is what then became known as *Kepler's third law of planetary motion*.

Kepler's three laws of planetary motion were developed empirically—meaning that they resulted from the analysis of accumulated observations rather than derived from a theory. No one, not even Kepler, knew the physical principals behind the laws of planetary motion at that time [1, 2]. It was not until the problem was tackled by the British mathematician Isaac Newton (1642–1727), many years after Kepler's death, that the reason was found. Newton found that each of Kepler's laws is a direct result of the universal law of gravitation [2].

Interestingly, the three Kepler's laws apply not only to planets, but to any orbiting object, including artificial satellites. That is, for a satellite orbiting the Earth (e.g., a GPS satellite), the orbital path takes the shape of an ellipse, with the Earth located at one of its focal points (Figure 3.3). As is well

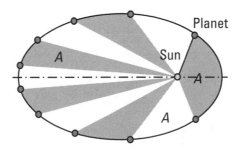

Figure 3.2 Illustration of Kepler's second law of planetary motion.

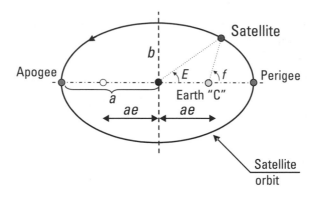

Figure 3.3 Geometry of satellite orbital ellipse.

known, an ellipse has two focal points with a separation proportional to the degree of ellipticity (i.e., highly elliptical orbits have wide separation between focal points, while near-circular orbits have very short separations). For a circular orbit, which is considered a special case of elliptical orbit, the two focal points merge to become one point—the center of the circle. The geometry of an ellipse can be totally described by two parameters, the semimajor axis of the satellite orbit (a) and the eccentricity of the satellite orbit (e). The latter parameter, e, describes the amount of deviation of the ellipse from a circle. The smaller the value of the eccentricity is, the shorter the separation between focal points, and vice versa. A satellite is closest to the Earth at a point called perigee, while it is farthest at a point called apogee (Figure 3.3). The other two Kepler's laws apply to satellite orbits in a similar manner to planets. If, for example, a particular orbital period is required for a particular satellite, the mean distance between the satellite and the Earth (i.e., the semimajor axis of the satellite orbit) must be determined in accordance with Kepler's third law of planetary motion. In the case of GPS, the orbital period is approximately 12 hours, which, based on Kepler's third law of planetary motion, corresponds to a satellite altitude of about 20,000 km above the Earth's surface.

3.2 Types of Orbits

Satellite orbits can vary, depending on, among other things, altitude, inclination with respect to the Earth's equatorial plane, and orbital period. Three different classes of satellite orbits are commonly used in practice, which are characterized by their altitudes, namely low Earth orbit (LEO), medium Earth orbit (MEO), and geostationary Earth orbit (GEO) [3]. LEOs are those satellite orbits of up to 2,000 km altitude, which are used for constellations of remote sensing satellites, altimeter satellites, and others. MEOs, on the other hand, are those satellite orbits of altitudes between 5,000 and 20,000 km. In fact, GPS belongs to this group of satellite orbits. Satellite orbits of altitude of 36,000 km are called GEOs, which are mainly used for communication purposes. A GEO circular orbit of zero inclination has a 24-hour period and appears to have a fixed location in the sky to an observer on the surface of the Earth. If, however, the orbital inclination is different from zero, the high-altitude (i.e., 36,000 km) circular orbit is called inclined geosynchronous orbit (IGSO), which does not appear to have a fixed location in the sky to an observer on the surface of the Earth. The ground track of an IGSO draws a large figure eight [3].

Other types of satellite orbits include highly elliptical and polar orbits. The former has a perigee (closest point to Earth) and an apogee (farthest point from Earth) at roughly 500 km and 50,000 km, respectively. Satellites with highly elliptical orbits provide communication services at high latitudes [3]. Polar orbits are those with an inclination of 90 degrees (i.e., perpendicular to the equator). They are fixed in space, and consequently each of them provides global coverage. The Transit system, the GPS predecessor, used circular, polar orbits with altitudes of about 1,100 km [4].

3.3 Ideal (Keplerian) Satellite Orbit

Satellite orbital motion results mainly from the Earth's gravitational attraction, with small contributions from celestial bodies (e.g., Sun and moon). If a satellite orbital motion obeys Kepler's laws of planetary motion, it is called Keplerian (or ideal). In fact, our discussion in Section 3.1, which states that a satellite orbital motion follows three Kepler's laws, is only true under certain ideal conditions. These conditions are: (1) all forces except the Earth's gravitational force are neglected; (2) the Earth's gravitational field is radially symmetric; (3) no atmosphere; and (4) satellite mass is negligibly small compared to the Earth's mass.

A Keplerian orbit can be fully described using only six parameters, known as Keplerian elements. Of these, only one changes with time, which describes the satellite position in the orbital ellipse and is known as the true anomaly f. The six Keplerian elements can be described as follows (see Figure 3.4):

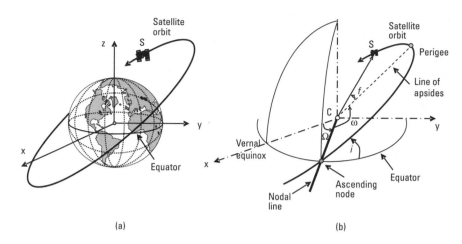

(a)   (b)

Figure 3.4 Keplerian elements. Note that (1) the nodal line is the line of intersection between the satellite orbital plane and the equatorial plane, and (2) line of apsides is the line connecting the perigee and apogee).

1. The semi-major axis of the satellite orbit (a);

2. Eccentricity of the satellite orbit (e);

3. *Right ascension of the ascending node* (Ω)—the angle between vernal equinox and nodal line, measured counterclockwise in the equatorial plane;

4. *Argument of perigee* (ω)—the angle between the nodal line and the line of apsides, measured counterclockwise (when looking toward the center of the Earth C) in the orbital plane;

5. The *inclination* (i) of the orbital plane (the half stretching from ascending node to descending node) with respect to the equatorial plane;

6. *True anomaly* (f)—the angle between the line of apsides and the line connecting the center of the Earth C with the satellite position S, measured counterclockwise (see Figure 3.4).

Two other types of anomalies are also used, namely eccentric anomaly (E) and mean anomaly (M), which are related to true anomaly. The geometrical relationship between the true and the eccentric anomalies is given in Figure 3.3. The mean anomaly, on the other hand, is the true anomaly corresponding to the motion of an imaginary satellite of a uniform angular velocity. As shown in Table 3.1, the mean anomaly, not the other two anomalies, is transmitted as part of the navigation message.

3.4 Perturbed Satellite Orbit

As indicated earlier, the assumption that a satellite orbit is Keplerian is only true under certain ideal conditions. In reality, however, satellite motion is not Keplerian, as small perturbations occur due to irregularities of Earth's gravitational field, attraction of celestial bodies, and other considerations. Nevertheless, acceleration due to geocentric gravitational attraction of the Earth is at least four orders of magnitude larger than perturbing accelerations (i.e., it determines the main shape of the satellite orbit).

In addition to geocentric gravitational attraction of the Earth, satellite motion is also affected by (see Figure 3.5):

- Noncentral attraction of the Earth (due mainly to the oblateness, or equatorial bulge, of the Earth and uneven lateral distribution of masses within the Earth).
- Gravitational attractions of the Sun, moon, and planets (known as third body effect).

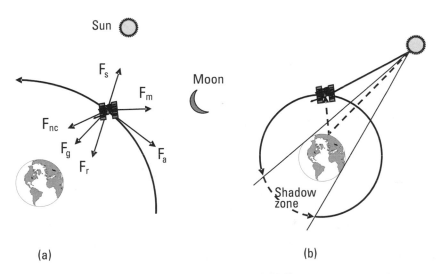

(a)   (b)

Figure 3.5 (a) Perturbing forces, and (b) solar radiation effect.

- Temporal variation of the gravity field due to tides (both ocean and solid Earth) and other deformation factors.

- Solar radiation pressure (both direct and albedo effects). Direct radiation pressure effect is caused by light photons emitted by the Sun, while indirect (known as albedo) effect is caused by solar radiation that arrives at the satellite after it has been reflected from the surface of the Earth—see Figure 3.5(b). Obviously the direct effect vanishes when the satellite is in the Earth's shadow zone.

- Atmospheric drag effect, which is caused by the friction between the satellite surface and particles of the surrounding atmosphere. It is particularly important for low-orbiting satellites (i.e., it is very small for GPS).

- Other factors include magnetic forces and solar wind.

The higher the satellite orbit is, the smaller the perturbations, and consequently the smoother the orbit. In the case of GPS, Keplerian orbit departs by as much as a few kilometers from the actual GPS orbit [4]. Such a departure is significant because, if neglected, it can cause considerable positioning error. In other words, Keplerian orbit can only be considered as a first approximation to the actual orbit.

3.5 GPS Broadcast Orbit

The GPS broadcast orbit, or broadcast ephemeris, is transmitted as part of the GPS navigation message in the form of *predicted* orbital parameters, which represent the perturbed GPS satellite orbit. At present (May 2006), a network of 12 monitoring stations collect GPS data on a continuous basis. The data is then transmitted to the GPS master control station, where it is fed into a Kalman filter to produce the orbital parameters [5]. A total of 16 parameters are needed to describe the perturbed orbit, although other information is also transmitted in each ephemeris record of a particular satellite (e.g., satellite clock correction parameters). Table 3.1 shows the 16-element broadcast orbit and their locations within an ephemeris record in the RINEX format (see Chapter 8 for information about the RINEX format). The first row contains the ephemeris reference time t_{0e}, rows 2–7 contain the six Keplerian elements, while the remaining parameters represent the perturbation parameters (nine elements). For the parameters in Table 3.1 to be fully understood, two parameters still need to be defined, namely the mean motion and argument of latitude. Mean motion, *n*, is the mean angular satellite velocity, which is a function of the orbital period only (or equivalently the semimajor axis, according to Kepler's third law). Mean motion difference, Δn (row 8 in Table 3.1), is a correction for the mean motion. Both the corrected mean motion and the mean anomaly (row 7 in Table 3.1) are used to compute the true anomaly. The argument of latitude, on the other hand, is the sum of the true anomaly and the argument of perigee.

After applying all of these corrections, a new set of parameters are obtained, which are used to compute the satellite cordinates at the time of signal transmission. The computed satellite coordinates will refer to the World Geodetic System of 1984 (WGS 84) system (see Appendix A for information on the WGS 84 system). It should be pointed out that the ephemeris records are updated every hour (most of the time). To ensure high positioning accuracy, only fresh ephemeris records (i.e., within the intended one-hour period) should be used. A user must examine two elements in the

Table 3.1
Sixteen-Element Broadcast Orbit and the Locations of the Elements Within an Ephemeris Record in the RINEX Format

No.	Parameter	Definition
1	t_{0e}	Ephemeris reference time (line 4)
2	$\sqrt{a}$	Square root of semimajor axis (line 3)
3	e	Eccentricity (line 3)
4	Ω_0	Right ascension parameter at t_{0e} (line 4)
5	ω	Argument of perigee (line 5)
6	i_0	Inclination at reference time t_{0e} (line 5)
7	M_0	Mean anomaly at reference time t_{0e} (line 2)
8	Δn	Mean motion difference (line 2)
9	Ω-dot	Rate of right ascension (line 5)
10	i-dot	Rate of inclination (line 6)
11	C_{rs}	Correction to orbital radius (line 2)
12	C_{rc}	Correction to orbital radius (line 5)
13	C_{us}	Correction to argument of latitude (line 3)
14	C_{uc}	Correction to argument of latitude (line 3)
15	C_{is}	Correction to inclination (line 4)
16	C_{ic}	Correction to inclination (line 4)

broadcast ephemeris, namely issue of data clock (IODC) and issue of data ephemeris (IODE), to detect any changes in the clock correction and ephemeris parameters, respectively (see Table 3.2). For completeness, Figure 3.6 represents an entire ephemeris record for one of the satellites in the RINEX format. An explanation of what each element in the record represents is given in Table 3.2.

```
 18 03  1  1 12  0  0.0-0.968435779214D-05 0.272848410532D-11 0.000000000000D+00
    0.108000000000D+03-0.804687500000D+02 0.476591280511D-08 0.117049746321D+01
   -0.418908894062D-05 0.343338993844D-02-0.307336449623D-06 0.515373415756D+04
    0.302400000000D+06-0.186264514923D-08-0.142655449866D+01 0.726431608200D-07
    0.963900001275D+00 0.387218750000D+03-0.309763566356D+01-0.854857036798D-08
   -0.201436962087D-09 0.100000000000D+01 0.119900000000D+04 0.000000000000D+00
    0.300000000000D+01 0.000000000000D+00-0.102445483208D-07 0.364000000000D+03
    0.298728000000D+06 0.000000000000D+00 0.000000000000D+00 0.000000000000D+00
```

Figure 3.6 An example ephemeris record for one PRN in the RINEX format.

Table 3.2
An Explanation of Parameters in Figure 3.6

Line No.	Parameter	Explanation (and Units)
Line 1	18	Satellite PRN number
	03 1 1 12 0 0.0	Epoch: t_{oc}, or time of clock
	−0.968435779214D−05	SV clock bias (sec)
	0.272848410532D−11	SV clock drift (sec/sec)
	0.000000000000D+00	SV clock drift rate (sec/sec^2)
Line 2	0.108000000000D+03	IODE
	−0.804687500000D+02	Correction to orbital radius (C_{rs}) (meters)
	0.476591280511D−08	Mean motion difference (Δn) (rad/sec)
	0.117049746321D+01	Mean anomaly (M_0) (radians)
Line 3	−0.418908894062D−05	Correction to argument of latitude (C_{uc}) (rad)
	0.343338993844D−02	Eccentricity (e)
	−0.307336449623D−06	Correction to argument of latitude (C_{us}) (rad)
	0.515373415756D+04	Square root of semimajor axis ($\sqrt{a}$) ($\sqrt{m}$)
Line 4	0.302400000000D+06	Ephemeris reference time (t_{0e}) (sec of GPS week)
	−0.186264514923D−08	Correction to inclination (C_{ic}) (rad)

Table 3.2 (continued)

Line No.	Parameter	Explanation (and Units)
	−0.142655449866D+01	Right ascension parameter (Ω_0) (rad)
	0.726431608200D−07	Correction to inclination (C_{is}) (rad)
Line 5	0.963900001275D+00	Inclination at reference time t_{0e} (i_0) (rad)
	0.387218750000D+03	Correction to orbital radius (C_{rc}) (meters)
	−0.309763566356D+01	Argument of perigee (ω) (rad.)
	−0.854857036798D−08	Rate of right ascension (rad/sec)
Line 6	−0.201436962087D−09	Rate of inclination (rad/sec)
	0.100000000000D+01	Codes on L2 Channel
	0.119900000000D+04	GPS Week number (to go with t_{0e})
	0.000000000000D+00	L2 P data flag
Line 7	0.300000000000D+01	SV accuracy (meters)
	0.000000000000D+00	SV health (usable if 0.0)
	−0.102445483208D−07	Differential group delay (T_{GD}) (seconds)
	0.364000000000D+03	IODC
Line 8	0.298728000000D+06	Transmission time of message (sec of GPS week)
	0.000000000000D+00	Fit interval (hours)—zero if not known
	0.000000000000D+00	Spare
	0.000000000000D+00	Spare

3.6   GPS Almanac

The GPS almanac is a subset of data containing information about the satellite orbit (coordinates) and satellite clock correction parameters. The almanac is transmitted as a part of the GPS navigation message. In comparison with the ephemeris, the almanac data is less accurate. The almanac parameters are updated every 6 days or less [6]. With the help of almanac data, a GPS receiver can rapidly acquire satellite signals [3]. GPS almanac is also used to predict satellite visibility at a particular location and time (see Section 3.7). Table 3.3 shows a complete almanac record in the YUMA format [7]. Both the almanac and ephemeris parameters can be downloaded free of charge over the Internet (e.g., from the U.S. Coast Guard Navigation Center [7] and the Crustal Dynamics Data Information System (CDDIS) anonymous FTP archive [8], respectively).

Table 3.3
Almanac Record for PRN-09 in GPS Week 325 (YUMA Format)

Parameter	Value
ID of PRN	09
Health status of satellite	000 (unusable if not 000)
Eccentricity (e)	0.1739120483E–001
Time of almanac generation t_a (sec. of week)	589824.0000
Orbital inclination i (rad)	0.9582130834
Rate of right ascension (rad/sec)	–0.7680319916E–008
Square root of semimajor axis ($\sqrt{a}$)	5153.612305
Right ascension parameter at weekly epoch (rad)	0.6326674561E+000
Argument of perigee (rad)	1.204548677
Mean anomaly (rad)	–0.6287860653E-001
Satellite clock bias $Af(0)$ (sec)	0.1440048218E–003
Satellite clock drift $Af(1)$ in seconds per seconds	0.5093170330E–010
GPS week	325

3.7   Satellite Visibility

Even under the full (official) constellation of 24 GPS satellites, there exist some periods of time where only 4 or 5 satellites are visible above a particular elevation angle, which may not be adequate for some GPS applications. Such a satellite visibility problem is expected more at high latitudes (higher than about 55 degrees) because of the nature of the GPS constellation [9]. This problem may also occur in some low- or mid-latitude areas for a particular period of time. For example, in urban and forested areas, the receiver's sky window is reduced as a result of the obstruction caused by the high-rise buildings and the trees. Because satellite geometry changes over time, the satellite visibility problem may be overcome by selecting a suitable observation time, which ensures a minimum number of visible satellites and/or a particular satellite geometry. To help users identify the best observation periods, GPS manufacturers have developed so-called mission-planning software packages, which predict the satellite visibility and geometry at any given location and time.

Mission-planning software predicts the satellite geometry based on the user's approximate location and the approximate satellite locations obtained from a recent almanac file for the GPS constellation. Mission-planning soft-ware provides a number of plots to help in planning the GPS survey or mission. The first plot is known as the sky plot, which represents the user's sky window by a series of concentric circles. Figure 3.7 shows the sky plot for Toronto on April 13, 2001, which was produced by the Ashtech Locus processor software. The center point represents the user's zenith, while the outer circle represents his or her horizon. Intermediate circles represent different elevation angles. The outer circle is also graduated from 0 degrees to 360 degrees to represent the satellite azimuth (direction) at any time. Once the user inputs his or her approximate location and the desired observation period, the path of each satellite in his or her view will be shown on the sky plot. This means that relative satellite locations, the satellite azimuth, and satellite elevation can be obtained. The user may also specify a certain elevation angle, normally 10 degrees or 15 degrees, to be used as a mask or cut-off angle. A mask angle is the angle below which the receiver will not track any satellite, even if the satellite is above the receiver's horizon.

Other important plots include the satellite availability plot, which shows the total number of visible satellites above the user-specified mask angle, and the satellite geometry plot, which is normally represented by the so-called dilution of precision parameters (see Chapter 4 for more details).

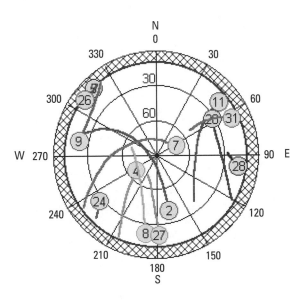

Figure 3.7 GPS sky plot for Toronto on April 13, 2001.

References

[1]   Chaisson, E., and S. McMillan, *Astronomy Today*, 4th ed., New Jersey: Prentice-Hall, 2002.

[2]   Wlaker, J. S., *Physics*, New Jersey: Prentice-Hall, 2002.

[3]   Seeber, G., *Satellite Geodesy*, second revised edition, Berlin, Germany: Walter de Gruyter, 2003.

[4]   Wells, D. E., et al., *Guide to GPS Positioning*, Fredericton, New Brunswick: Canadian GPS Associates, 1987.

[5]   Global View, "NGA Data Drives Improved Accuracy," *GPS World*, Vol. 16, No. 10, October 2005, p. 14.

[6]   ICD-GPS-200C, "Interface Control Document ICD-GPS-200, Revision C," IRN-200C-005R1, ARINC Research Corporation, El Segundo, CA, January 14, 2003, http://www.navcen.uscg.gov/gps/modernization/default.htm.

[7]   U.S. Coast Guard Navigation Center, accessed November 2005, http://www.navcen.uscg.gov/gps/almanacs.htm.

[8]   Crustal Dynamics Data Information System (CDDIS), accessed November 2005, ftp://cddis.gsfc.nasa.gov/gps/products.

[9]   El-Rabbany, A., *The Effect of Physical Correlations on the Ambiguity Resolution and Accuracy Estimation in GPS Differential Positioning*, Technical Report No. 170, Department of Geodesy and Geomatics Engineering, Fredericton, New Brunswick, Canada: University of New Brunswick, 1994.

4

GPS Errors and Biases

GPS measurements are affected by several types of random errors and biases (systematic errors). These errors may be classified as those originating at the satellites, those originating at the receiver, and those that are due to signal propagation (atmospheric refraction) [1]. Figure 4.1 shows the main errors and biases.

The errors originating at the satellites include ephemeris, or orbital, errors, satellite clock errors, and the effect of selective availability. The latter was intentionally implemented by the U.S. DoD to degrade the autonomous

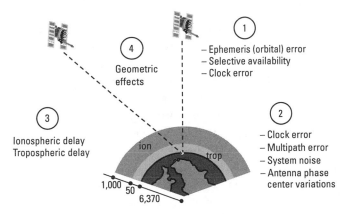

Figure 4.1 Main GPS errors and biases.

GPS accuracy for security reasons. It was, however, switched off on May 2, 2000 [2]. The errors originating at the receiver include receiver clock errors, multipath error, receiver noise, and antenna phase center variations. The signal propagation errors include the delays of the GPS signal as it passes through the atmospheric layers (mainly the ionosphere and the troposphere). In fact, it is only in a vacuum (free space) that the GPS signal travels, or propagates, at the speed of light.

In addition to the effect of these errors, the accuracy of the computed GPS position is also affected by the geometric locations of the GPS satellites as seen by the receiver. The more spread out the satellites are in the sky, the better the accuracy obtained (Figure 4.1).

As shown in Chapter 2, some of these errors and biases can be eliminated or reduced through appropriate combinations of the GPS observables. For example, combining L1 and L2 observables removes, to a high degree of accuracy, the effect of the ionosphere. Mathematical modeling of these errors and biases is also possible. In this chapter, the main GPS error sources are introduced together with the methods for their treatment.

4.1   GPS Ephemeris Errors

Satellite positions as a function of time, which are included in the broadcast satellite navigation message, are predicted from previous GPS observations at the monitoring stations. Typically, overlapping 4-hour GPS data spans are used by the operational control system to predict fresh satellite orbital elements for each 1-hour period. As might be expected, modeling the forces acting on the GPS satellites will not in general be perfect, which causes some errors in the estimated satellite positions, known as ephemeris errors. Until recently, the ephemeris error was on the order of about 2m. However, with the recent addition of six new monitoring stations as part of the control segment modernization program (see Chapter 2), the GPS ephemeris error is reported to be in the order of 1.6m (i.e., about 20 percent accuracy improvement [3]).

An ephemeris error for a particular satellite is identical to all GPS users worldwide [4]. However, as different users see the same satellite at different view angles, the effect of the ephemeris error on the range measurement, and consequently on the computed position, is different. This means that combining (differencing) the measurements of two receivers simultaneously tracking a particular satellite cannot totally remove the ephemeris error. However, users at short separations (baselines) will have an almost identical range error due to the ephemeris error, which can essentially be removed through differencing the observations. For relative positioning (see Chapter 5), the following general rule gives a rough estimate of the effect of the ephemeris error on the baseline solution: the baseline error / the baseline length = the satellite position error / the range satellite [5]. This means that if the satellite position error is 1m and the baseline length is 100 km, then the expected baseline error due to ephemeris error is approximately 5 mm.

Some applications, such as studies of the crustal dynamics of the Earth, require more precise ephemeris data than the broadcast ephemeris. The PPP technique (see Chapter 5) has recently evolved, which also requires the availability of precise ephemeris (as well as satellite clock corrections). To support these applications, several institutions, such as the International GNSS Service (IGS), formerly International GPS Service, have developed post-mission precise orbital service. Precise ephemeris data is based on GPS data collected at a global GPS network coordinated by the IGS. At present, IGS precise ephemeris data is available to users with some delay, which varies from 3 hours for the observed half of the IGS ultrarapid orbit to 17 hours for the rapid orbit, to about 13 days for the most precise IGS final orbit [3]. The three types of precise orbits have accuracies better than 5 cm. However, they differ in the accuracy of the satellite clock corrections (see Section 3.3). Users requiring real-time precise orbit data can use the predicted half of the IGS ultrarapid orbit data, which are accurate to about 10 cm. In fact, some organizations such as Natural Resources Canada (NRCan) and the Jet Propulsion Laboratory (JPL) now offer real-time precise orbit data (and satellite clock corrections) over the Internet. Further information on precise orbit data services is given in Chapter 7.

4.2 Selective Availability

GPS was originally designed so that real-time autonomous positioning and navigation with the civilian C/A-code receivers would be less precise than military P-code receivers. Surprisingly, the obtained accuracy was almost the same from both receivers [6]. To ensure national security, the U.S. DoD implemented the so-called selective availability (SA) on Block II GPS satellites to deny accurate real-time autonomous positioning to unauthorized users. SA was officially activated on March 25, 1990 [7].

SA introduces two types of errors [6]. The first one, called delta error, results from dithering the satellite clock and is common to all users worldwide. The second one, called epsilon error, is an additional slowly varying orbital error. With SA turned on, nominal horizontal and vertical errors can be up to 100m and 156m, respectively, at the 95 percent probability level. Figure 4.2 shows how the horizontal position of a stationary GPS receiver varies over time, mainly as a result of the effect of SA. Like the range error due to ephemeris error, the range error due to epsilon error is almost identical between users at short separation distances. Therefore, using differential GPS (DGPS), which is discussed in greater detail in Chapter 5, overcomes the effect of the epsilon error. In fact, DGPS provided better real-time positioning accuracy than the stand-alone P-code receiver due to the elimination or the reduction of the common errors, including SA [4]. This is particularly true for short baselines.

Following extensive studies, the U.S. government deactivated SA on May 2, 2000, resulting in a much-improved autonomous GPS accuracy [2]. With the SA switched off, the autonomous GPS accuracy was improved by a factor of seven or more. Figure 4.3 shows the GPS errors after SA was turned off. In fact, the deactivation of SA has opened the door for faster growth of GPS markets (e.g., vehicle navigation and enhanced-911). Although the deactivation of SA would have little impact on the DGPS accuracy, it would reduce the cost of installing and operating a DGPS. This is mainly because of the reduction in the required transmission rate.

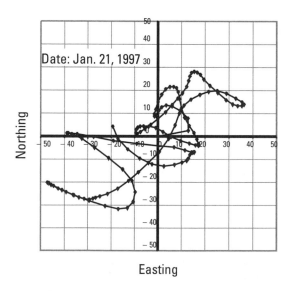

Easting

Figure 4.2 Position variation of a stationary GPS receiver due to SA.

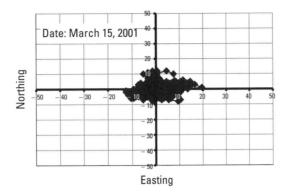

Easting

Figure 4.3 Position variation of a stationary GPS receiver after deactivating SA.

4.3 Satellite and Receiver Clock Errors

Each GPS Block II and Block IIA satellite contains four atomic clocks: two cesium and two rubidium [8]. The newer generation Block IIR satellites carry three rubidium clocks only [9]. One of the onboard clocks, primarily a cesium for Block II and IIA, is selected to provide the frequency and the timing requirements for generating the GPS signals. The others are backups [8].

The GPS satellite clocks, although highly accurate, are not perfect. Their stability is about 1 to 2 parts in 10^{13} or better over a period of one day. This means that the satellite clock error is about 8.64 to 17.28 ns per day. The corresponding range error is 2.59m to 5.18m, which can be easily calculated by multiplying the clock error by the speed of light (i.e., 299,729,458 m/s). Cesium clocks tend to behave better over a longer period of time compared with rubidium clocks. In fact, the stability of the cesium clocks over a period of 10 or more days improves to several parts in 10^{14} [8]. The performance of the satellite clocks is monitored by the ground control system. The offset between the satellite clock and GPS time is continually monitored at the MCS and transmitted as a part of the navigation message in the form of three coefficients of a second-degree polynomial, representing satellite clock bias (seconds), drift (sec/sec), and drift rate(sec/sec^2), respectively [7, 10].

Satellite clock errors cause additional errors to the GPS measurements. These errors are common to all users observing the same satellite and can effectively be removed through differencing between the receivers. Applying the satellite clock correction in the navigation message can also correct the satellite clock errors. This, however, leaves an error of the order of 7 ns, which translates to a range error of 2.1m (one nanosecond error is equivalent to a range error of about 30 cm) [4].

GPS receivers, in contrast, use inexpensive crystal clocks, which are much less accurate than the satellite clocks [1]. As such, the receiver clock error is much larger than that of the GPS satellite clock. It can, however, be removed through differencing between the satellites, or it can be treated as an additional unknown parameter in the estimation process. Precise external clocks (usually cesium or rubidium) are used in some applications instead of the internal receiver clock. Some IGS reference stations use external hydrogen masers; the most stable of the atomic clocks (several parts in 10^{15}) at the present time [11]. Although the external atomic clocks have superior performance compared with the internal receiver clocks, they are expensive and bulky.

4.4 Multipath Error

Multipath is a major error source for both the carrier-phase and pseudorange measurements. Multipath errors occur when the GPS signal arrives at the receiver antenna through different paths [5]. These paths can be the direct line-of-sight signal and reflected signals from objects surrounding the receiver antenna (Figure 4.4).

Multipath distorts the original signal through interference with the reflected signals at the GPS antenna. It affects both the carrier-phase and pseudorange measurements; however, its size is much larger in the pseudorange measurements. The size of the carrier-phase multipath can reach a maximum value of a quarter of a cycle (about 4.8 cm for the L1 carrier phase). The pseudorange multipath can theoretically reach several tens of meters for the C/A-code measurements. However, with new advances in signal processing and receiver technology, actual pseudorange multipath is reduced dramatically. Examples of such technologies include narrow correlator, multipath estimating delay lock loop, and strobe techniques [12].

Under the same environment, the presence of multipath errors can be verified using a day-to-day correlation (or similarity) of the estimated residuals. This is because the satellite-reflector-antenna geometry repeats every sidereal day. However, multipath errors in the undifferenced pseudorange measurements can be identified if dual-frequency observations are available. A good general multipath model is still not available, mainly because of the variant satellite-reflector-antenna geometry. There are, however, several other options to reduce the effect of multipath. The straightforward option is

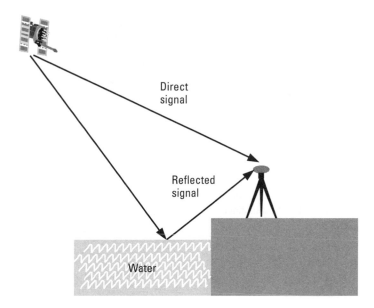

Figure 4.4 Multipath effect.

to select an observation site with no reflecting objects in the vicinity of the receiver antenna. Another option to reduce the effect of multipath is to use a receiver that takes advantage of multipath mitigation techniques. The use of a chock-ring antenna (a chock-ring device is a ground plane that has several concentric metal hoops, which attenuate the reflected signals) can also reduce the effect of multipath (see Figure 4.5). As the GPS signal is right-handed circularly polarized while the reflected signal is left-handed, reducing the effect of multipath may also be achieved by using an antenna with a matching polarization to the GPS signal (i.e., right-handed). The disadvantage of this option, however, is that the polarization of the multipath signal becomes right-handed again if it is reflected twice [13].

4.5 Antenna Phase Center Variation

As stated in Chapter 2, a GPS antenna receives the incoming satellite signal and then converts its energy into an electric current, which can be handled by the GPS receiver [14]. The point at which the GPS signal is received is called the antenna phase center [7]. Generally, the antenna phase center does not coincide with the physical (geometrical) center of the antenna. It varies depending on the elevation and the azimuth of the GPS satellite as well as the

Figure 4.5 A chock-ring antenna (*courtesy of:* Thales Navigation; available at http://products.thalesnavigation.com/en/products/product.asp?PRODID=60).

intensity of the observed signal. As a result, an additional range error can be expected [7].

The size of the error caused by the antenna-phase-center variation depends on the antenna type and is typically on the order of a few centimeters. It is, however, difficult to model the antenna-phase-center variation, and therefore care has to be taken when selecting the antenna type [1]. For short baselines with the same types of antennas at each end, the phase-center error can be canceled if the antennas are oriented in the same direction [15]. Mixing different types of antennas or using different orientations will not cancel the error. Due to its rather small size, this error is neglected in most of the practical GPS applications.

It should be pointed out that phase-center errors could be different on L1 and L2 carrier-phase observations. This can affect the accuracy of the ion-osphere-free linear combination, particularly when observing short baselines. As mentioned before, for short baselines, the errors are highly correlated (i.e., highly similar) over distance and cancel sufficiently through differencing. Therefore, single frequency measurements might be more appropriate for short baselines in the static mode (see Chapter 5 for details on the static GPS positioning mode).

4.6 Receiver Measurement Noise

The receiver measurement noise results from the limitations of the receiver's electronics. A good GPS (i.e., receiver and antenna) should have a minimum noise level. Generally, a GPS receiver performs a self-test when the user turns it on. However, for expensive precise GPSs, it might be important for the user to perform the system evaluation. Two tests can be performed to evaluate GPS receivers: zero baseline and short baseline [16].

A zero baseline test is used to evaluate the receiver performance. The test involves using one antenna/preamplifier followed by a signal splitter that feeds two or more GPS receivers (see Figure 4.6). Several receiver problems such as interchannel biases and cycle slips can be detected with this test. As one antenna is used, the baseline solution should be zero. In other words, any nonzero value is attributed to the receiver noise. Although the zero baseline test provides useful information on the receiver performance, it does not provide any information on the antenna/preamplifier noise. The contribution of the receiver measurement noise to the range error will depend very much on the quality of the GPS receiver.

To evaluate the actual field performance of a GPS, it is necessary to include the antenna/preamplifier noise component [16]. This can be done using short baselines of a few meters apart, observed on two consecutive days (see Figure 4.7). In this case, the double difference residuals of one day would contain the system noise and the multipath effect. All other errors would cancel sufficiently. As the multipath signature repeats every sidereal day, differencing the double difference residuals between the two consecutive days eliminates the effect of multipath and leaves only the system noise.

4.7 Ionospheric Delay

At the uppermost part of the Earth's atmosphere, ultraviolet and X-ray radiations coming from the Sun interact with the gas molecules and atoms. These interactions result in *gas ionization*: a large number of free *negatively charged*

Figure 4.6 Zero baseline test for evaluating the performance of a GPS receiver.

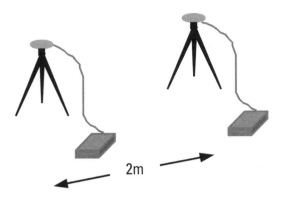

Figure 4.7 Short baseline test for evaluating the performance of a GPS.

electrons and *positively charged* atoms and molecules [17]. This region of the atmosphere where gas ionization takes place is called the ionosphere. It extends from an altitude of approximately 50 km to about 1,000 km or even more (see Figure 4.1). In fact, the upper limit of the ionospheric region is not clearly defined [18, 19].

The electron density within the ionospheric region is not constant; it changes with altitude. As such, the ionospheric region is divided into subregions, or layers, according to the electron density. These layers are named D (50–90 km), E (90–140 km), F1 (140–210 km), and F2 (210–1,000 km), with F2 usually being the layer of maximum electron density. The altitude and thicknesses of these layers vary with time, as a result of the changes in the Sun's radiation and the Earth's magnetic field. For example, the F1 layer disappears during the night and is more pronounced in the summer than in the winter [18].

The question that arises is: how does the ionosphere affect GPS measurements? The ionosphere is a dispersive medium, which means it bends the GPS radio signal and changes its speed as it passes through the various ionospheric layers to reach a GPS receiver. Bending the GPS signal path causes a negligible range error, particularly if the satellite elevation angle is greater than 5 degrees. It is the change in the propagation speed that causes a significant range error and therefore should be accounted for. The ionosphere speeds up the propagation of the carrier phase beyond the speed of light, while it slows down the PRN code (and the navigation message) by the same amount. That is, the receiver-satellite distance will be too short if measured by the carrier phase and too long if measured by the code, compared with the actual distance [7]. The ionospheric delay is proportional to the number of free electrons along the GPS signal path, called the total electron content (TEC). TEC, however, depends on a number of factors [20]:

1. *Time of day.* Electron density reaches a daily maximum early in the afternoon and a minimum after midnight at local time. This ionospheric cycle reflects the diurnal rotation of the Earth.

2. *Time of year.* Electron density is higher in winter than in summer. This seasonal ionospheric cycle reflects the annual rotation of the Earth around the Sun.

3. *The 11-year solar cycle.* Electron density reaches a maximum value every approximately 11 years, which corresponds to a peak in the solar flare activities known as the solar cycle peak—a solar flare is a sudden release of large amounts of energy by the sun in the solar atmosphere. Figure 4.8 shows high and low solar flare activities. An interesting phenomenon, which has been observed and well documented for over 300 years, is that during solar maximum, a large number of dark spots, known as sunspots, appear on the surface of the Sun. During solar minimum, however, the number of sunspots is lowest. Figure 4.9 represents the 11-year solar cycle of daily sunspot numbers since 1960. As can be seen from Figure 4.9, we are currently (at the time of writing in 2006) approaching the solar minimum (end) of cycle number 23, which is expected in 2006–2007.

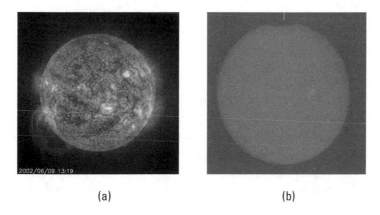

(a)   (b)

Figure 4.8 (a) Solar flare at high solar activities; (b) a solar minimum (quiet Sun). (*Source:* http://www.sec.noaa.gov/ImageGallery; accessed November 17, 2005.)

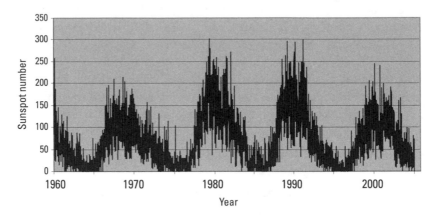

Figure 4.9 The 11-year solar cycle of daily sunspot numbers since 1960. (Source: ftp://ftp.ngdc.noaa.gov/STP/SOLAR_DATA/SUNSPOT_NUMBERS/RIDAILY.PLT.)

4. *Geographic location.* Electron density is lowest in the midlatitude regions and highly irregular in the polar, auroral, and equatorial regions.

As the ionosphere is a dispersive medium, it causes a delay that is frequency-dependent. The lower the frequency is, the greater the delay will be (i.e., the L2 ionospheric delay is greater than that of L1). Generally, ionospheric delay is of the order of 5m to 15m, but can reach over 150m under extreme solar activities, at midday, and near the horizon [5].

This discussion shows that the electron density in the ionosphere varies with time and location. It is, however, highly correlated over relatively short distances, and therefore differencing the GPS observations between users of short separation can remove the major part of the ionospheric delay. Taking advantage of the ionosphere's dispersive nature, the ionospheric delay can be determined with a high degree of accuracy by combining the P-code pseudorange measurements on both L1 and L2. Unfortunately, however, the P-code is accessible by authorized users only. With the addition of a second C/A-code on L2 as part of the modernization program, this limitation will be removed [2]. The L1 and L2 carrier-phase measurements may be combined in a similar fashion to determine the variation in the ionospheric delay, rather than the absolute value. Users with dual-frequency receivers can combine the L1 and L2 carrier-phase measurements to generate the ionosphere-free linear combination to largely (up to a few centimeters) remove the ionospheric delay [20]. The disadvantages of the ionosphere-free linear combination, however, are: (1) it has a relatively higher observation noise, and (2) it does not preserve the integer nature of the ambiguity parameters. As such, the ionosphere-free linear combination is not recommended for short baselines. Single-frequency users cannot take advantage of the dispersive nature of the ionosphere. They can, however, use one of the empirical ionospheric models to correct up to 60 percent of the delay [17]. The most widely used model is the Klobuchar model, the coefficients of which are transmitted as part of the navigation message. Another solution for users with single-frequency GPS receivers is to use corrections from one of the regional or global ionospheric services offered by organizations such as NOAA and IGS.

4.8 Tropospheric Delay

The troposphere is the electrically neutral atmospheric region that extends up to about 50 km from the surface of the Earth (see Figure 4.1). The troposphere is a nondispersive medium for radio frequencies below 15 GHz [21]. As a result, it delays the GPS carriers and codes identically. That is, the measured satellite-to-receiver range will be longer than the actual geometric range, which means that a distance between two receivers will be longer than the actual distance. Unlike the ionospheric delay, the tropospheric delay cannot be removed by combining the L1 and the L2 observations. This is mainly because the tropospheric delay is frequency-independent.

The tropospheric delay depends on the temperature, pressure, and humidity along the signal path through the troposphere. Signals from satellites at low elevation angles travel a longer path through the troposphere than those at higher elevation angles. Therefore, the tropospheric delay is minimized at the user's zenith and maximized near the horizon. Tropospheric delay results in values of about 2.3m at zenith (satellite directly overhead), about 9.3m for a 15-degree elevation angle, and about 20–28m for a 5-degree elevation angle [22].

Tropospheric delay may be broken into two components, dry (also called hydrostatic) and wet. The dry component represents about 90 percent of the delay and can be predicted to a high degree of accuracy using mathematical models (e.g., the Saastamoinen model) [23]. The wet component of the tropospheric delay depends on the water vapor along the GPS signal path. Unlike the dry component, the wet component is not easy to predict. Several mathematical models use surface meteorological measurements (atmospheric pressure, temperature, and partial water vapor pressure) to compute the wet component. Unfortunately, however, the wet component is weakly correlated with surface meteorological data, which limits its prediction accuracy. Recently, a number of regional and local monitoring networks have been established to generate tropospheric corrections. Among them is the NOAA tropospheric correction model, which incorporates GPS observations into numerical weather prediction models [24]. The model is distributed as a two-dimensional grid file, which is updated hourly and contains the zenith tropospheric delay (current and forecast) over the United States and surrounding regions (including a large portion of Canada). Future expansion of the model will include Alaska, all of Canada, and other North American regions [25].

4.9   Other Errors and Biases

In GPS relative positioning, some errors and biases are usually neglected, as they are either canceled out or reduced to a negligible level through differencing. In PPP, however, all errors and biases must be modeled with sufficient accuracy. As such, in addition to the ones discussed earlier, other errors and biases such as satellite antenna offsets, phase wind-up, site displacement effects, and equipment (hardware) delays must be considered in undifferenced PPP. This section provides a brief treatment of these additional errors and biases.

Satellite antenna offset (also known as satellite attitude error) refers to the separation between the satellite center of mass and its antenna phase center. As is well known, the GPS measurements are made with respect to the satellite antenna phase center. However, while the broadcast ephemeredes refer to the satellite antenna phase center, precise orbit and clock products (from IGS, for example) refer to the satellite center of mass. Satellite antenna offsets are calibrated and can be obtained from, for example, the IGS Web site.

Phase windup refers to the change in the measured phase as a result of mutual orientation of the satellite and receiver antennas. In general, satellite antennas experience two types of rotations while orbiting the Earth: (1) a slow rotation as a result of directing the satellite's solar panels toward the Sun, and (2) a rapid rotation (up to one revolution in less than 30 minutes) during eclipsing season, which results from reorienting the satellite's solar panels toward the Sun [26]. As shown by [23], a complete antenna rotation (i.e., a change in the antenna azimuth by 360 degrees) causes a change in the measured phase by one wavelength. In addition to PPP, phase windup affects relative positioning over large distances (e.g., over several hundred kilometers). Phase windup can be accounted for using mathematical models [26].

Site displacement effect represents periodic movements of the GPS site (station), which results mainly from solid Earth tides, ocean loading, and effect of polar motion. Solid Earth tide is the periodic movement of the solid Earth as a result of gravitational attractions from celestial bodies, particularly the Moon and the Sun. In fact, these are the same gravitational attractions that cause the ocean tides. The second effect, ocean loading, represents the deformation caused by the load of the ocean tides on sites that are close to the coasts. Sites that are far from the ocean have negligible ocean loading effect [26]. Polar motion represents the small changes in the locations of the poles, which in turn changes the Earth's rotational axis. As a result, locations of points on the Earth's crust with respect the Earth's axis of rotation will change, which causes small changes to the Earth's centrifugal acceleration at those point. This in turns causes periodic site deformation, which can be accounted for using well-known mathematical models [26].

Hardware delay (also called equipment or instrument delay) represents the time delays that occur at the satellite and the receiver. Satellite hardware delay is the time difference between signal generation (by the signal generator in the satellite) and signal transmission (by the satellite antenna). Receiver hardware delay, on the other hand, is the time difference between signal reception (by the GPS receiver antenna) and the signal processing inside the GPS receiver (i.e., receiver signal correlator). Unfortunately, hardware delays are frequency-dependent (i.e., the delay on L1 is different from that on L2). As well, the pseudorange hardware delay is different from that of the carrier phase [20]. The difference between the delays on L1 and L2 is called interfrequency (or differential) bias or equipment delay. Fortunately, hardware delays can be calibrated through GPS observations collected at reference stations with precisely known coordinates. It should be pointed out that the IGS products are consistent with P-code observations rather than C/A-code observations. As such, single-frequency users of C/A-code observations will have to apply the "C1-P1" differential biases, which can be obtained from the IGS Web site [27].

4.10 Satellite Geometry Measures

The various types of errors and biases discussed earlier directly affect the accuracy of the computed GPS position. Proper modeling of those errors and biases, and appropriate combinations of the GPS observables, will improve the positioning accuracy. However, these are not the only factors that affect the resulting GPS accuracy. Satellite geometry, which represents the geometric locations of the GPS satellites as seen by the receiver(s), plays a very important role in the total positioning accuracy [5]. The stronger the satellite geometry strength is, the higher the positioning accuracy obtained will be. As such, the overall positioning accuracy of GPS is measured by the combined effect of the unmodeled measurement errors and the effect of the satellite geometry.

Good satellite geometry is obtained when the satellites are spread out in the sky [28]. In general, the more spread out the satellites are in the sky, the better the satellite geometry will be, and vice versa. Figure 4.10 shows a simple graphical explanation of the satellite geometry effect using two satellites (assuming a two-dimensional case). In such a case, the receiver will be located at the intersection of two arcs of circles; each has a radius equal to the receiver-satellite distance and a center at the satellite itself. Because of the measurement errors, the measured receiver-satellite distance will not be exact, and an uncertainty region on both sides of the estimated distance will be present. Combining the measurements from the two satellites, it can be seen that the receiver will in fact be located somewhere within the uncertainty area, the hatched area. It is known from statistics that, for a certain probability level, if the size of the uncertainty area is small, the computed receiver's position will be precise. As shown in Figure 4.10(a), if the two satellites are far apart (i.e., spread out), the size of the uncertainty area will be

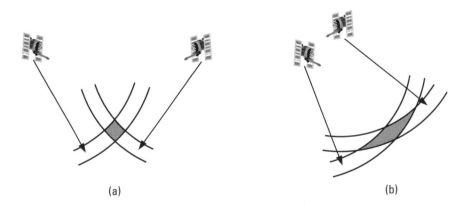

(a)   (b)

Figure 4.10 (a) Good satellite geometry; (b) bad satellite geometry.

small, resulting in good satellite geometry. Similarly, if the two satellites are close to each other, as in Figure 4.10(b), the size of the uncertainty area will be large, resulting in poor satellite geometry.

The satellite geometry effect can be measured by a single dimensionless number called the dilution of precision (DOP). The lower the value of the DOP is, the better the geometric strength will be, and vice versa [7, 10]. The DOP number is computed based on the relative receiver-satellite geometry at any instance (i.e., it requires the availability of both the receiver and the satellite coordinates). Approximate values for the coordinates are generally sufficient though, which means that the DOP value can be determined without making any measurements. As a result of the relative motion of the satellites and the receiver(s), the value of the DOP will change over time. The changes in the DOP value, however, will generally be slow except in the following two cases: (1) a satellite is rising or falling as seen by the user's receiver, and (2) there is an obstruction between the receiver and the satellite (e.g., when passing under a bridge).

In practice, various DOP forms are used, depending on the user's need [28]. For example, for the general GPS positioning purposes, a user may be interested in examining the effect of the satellite geometry on the quality of the resulting three-dimensional position (latitude, longitude, and height). This could be done by examining the value of the position dilution of precision (PDOP). In other words, PDOP represents the contribution of the satellite geometry to the three-dimensional positioning precision. PDOP can be broken into two components: horizontal dilution of precision (HDOP) and vertical dilution of precision (VDOP). The former represents the satellite geometry effect on the horizontal component of the positioning accuracy, while the latter represents the satellite geometry effect on the vertical component of the positioning accuracy. Because a GPS user can track only those satellites above the horizon, VDOP will *always* be larger than HDOP. As a result, the GPS height solution is expected to be less precise than the horizontal solution. The VDOP value could be improved by supplementing GPS with other sensors, for example, the pseudolites (see Chapter 9 for details). Other commonly used DOP forms include the time dilution of precision (TDOP) and the geometric dilution of precision (GDOP). GDOP represents the combined effect of the PDOP and the TDOP.

To ensure high-precision GPS positioning, a PDOP value of five or less is usually recommended. In fact, the actual PDOP value is usually much less than five, with a typical average value in the neighborhood of two. As mentioned in Chapter 3, most GPS software packages have the ability to predict the satellite geometry based on the user's approximate location and the approximate satellite locations obtained from a recent almanac file for the GPS constellation. The satellite geometry plot is normally represented by the PDOP, HDOP, and VDOP. Figure 4.11 shows the satellite availability and geometry for Toronto on April 13, 2001, which was produced by the Ashtech Locus processor software.

4.11 User Equivalent Range Error

It has been shown that the GPS positioning accuracy is measured by the combined effect of the unmodeled measurement errors and the satellite geometry. The unmodeled measurement errors will certainly be different from one satellite to another, mainly because of the various view angles. In addition, the ranging errors for the various satellites will have a certain degree of similarity (i.e., correlation). To rigorously determine the expected GPS positioning accuracy, we may apply an estimation technique such as the least-squares method [23]. The least-squares method estimates the user's position (location) as well as its covariance matrix. The latter tells us how well the user's position is determined. In fact, the covariance matrix reflects the combined effect of the measurement errors and the satellite geometry.

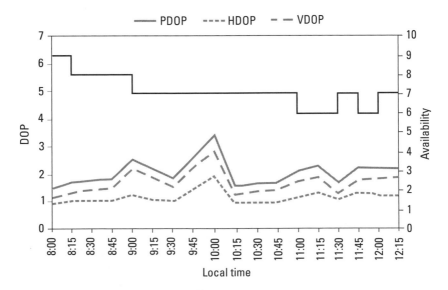

Figure 4.11 Satellite availability and geometry.

A more simplified way of examining the GPS positioning accuracy may be achieved through the introduction of the user equivalent range error (UERE). Assuming that the measurement errors for all the satellites are identical and independent, a quantity known as the UERE may be defined as the root-sum-square of the various errors and biases discussed earlier [7]. Multiplying the UERE by the appropriate DOP value produces the expected precision of the GPS positioning at the one-sigma (1-σ) level (see Appendix B). To obtain the precision at the 2-σ level, sometimes referred to as at least 95 percent of the time, we multiply the results by a factor of two (see Appendix B). For example, assuming that the UERE is 5m for the standalone GPS receiver, and taking a typical value of HDOP as 1.5, then the 95 percent positional accuracy will be 5 × 1.5 × 2 = 15m.

References

[1] Kleusberg, A., and R. B. Langley, "The Limitations of GPS," *GPS World*, Vol. 1, No. 2, March/April 1990, pp. 50–52.

[2] Shaw, M., K. Sandhoo, and D. Turner, "Modernization of the Global Positioning System," *GPS World*, Vol. 11, No. 9, September 2000, pp. 36–44.

[3] IGS, "IGS Product Table," http://igscb.jpl.nasa.gov/components/prods.html.

[4] El-Rabbany, A., *The Effect of Physical Correlations on the Ambiguity Resolution and Accuracy Estimation in GPS Differential Positioning*, Technical Report No. 170, Department of Geodesy and Geomatics Engineering, University of New Brunswick, Fredericton, New Brunswick, Canada, 1994.

[5] Wells, D. E., et al., *Guide to GPS Positioning*, Fredericton, New Brunswick: Canadian GPS Associates, 1987.

[6] Georgiadou, Y., and K. D. Doucet, "The Issue of Selective Availability," *GPS World*, Vol. 1, No. 5, September/October 1990, pp. 53–56.

[7] Hoffmann-Wellenhof, B., H. Lichtenegger, and J. Collins, *Global Positioning System: Theory and Practice*, 3rd ed., New York: Springer-Verlag, 1994.

[8] Langley, R. B., "Time, Clocks, and GPS," *GPS World*, Vol. 2, No. 10, November/December 1991, pp. 38–42.

[9] U.S. Naval Observatory, "USNO GPS Time Transfer," http://tycho.usno.navy.mil/gpstt.html.

[10] Kaplan, E., *Understanding GPS: Principles and Applications*, Norwood, MA: Artech House, 1996.

[11] Misra, P., and P. Enge, *Global Positioning System: Signals, Measurements and Performance*, Lincoln, MA: Ganga-Jamura Press, 2001

[12] Linyuan X., "Multipath in GPS Navigation and Positioning," *GPS Solutions*, Vol. 8, 2004, pp. 49–50.

[13] Weill, L. R., "Conquering Multipath: The GPS Accuracy Battle," *GPS World*, Vol. 8, No. 4, April 1997, pp. 59–66.

[14] Langley, R. B., "The GPS Receiver: An Introduction," *GPS World*, Vol. 2, No. 1, January 1991, pp. 50–53.

[15] Schupler, B. R., and T. A. Clark, "How Different Antennas Affect the GPS Observable," *GPS World*, Vol. 2, No. 10, November/December 1991, pp. 32–36.

[16] Nolan, J., S. Gourevitch, and J. Ladd, "Geodetic Processing Using Full Dual Band Observables," *Proc. ION GPS-92, 5th Intl. Technical Meeting*, Satellite Div., Institute of Navigation, Albuquerque, NM, September 16–18, 1992, pp. 1033–1041.

[17] Klobuchar, J. A., "Ionospheric Effects on GPS," *GPS World*, Vol. 2, No. 4, April 1991, pp. 48–51.

[18] Komjathy, A., *Global Ionospheric Total Electron Content Mapping Using the Global Positioning System,* Ph.D. dissertation, Department of Geodesy and Geomatics Engineering, Technical Report No. 188, University of New Brunswick, Fredericton, New Brunswick, Canada, 1997.

[19] Langley, R. B., "GPS, the Ionosphere, and the Solar Maximum," *GPS World*, Vol. 11, No. 7, July 2000, pp. 44–49.

[20] Teunissen, P. J. G. and A. Kleusberg (eds.), *GPS for Geodesy*, 2nd Ed., New York: Springer-Verlag, 1998.

[21] Hay, C., and J. Wong, "Enhancing GPS: Tropospheric Delay Prediction at the Master Control Station," *GPS World*, Vol. 11, No. 1, January 2000, pp. 56–62.

[22] Brunner, F. K., and W. M. Welsch, "Effect of the Troposphere on GPS Measurements," *GPS World*, Vol. 4, No. 1, January 1993, pp. 42–51.

[23] Leick, A., *GPS Satellite Surveying*. 3rd Ed., New Jersey: John Wiley & Sons, 2004.

[24] Gutman, S., T. Fuller-Rowell, and D. Robinson, "Using NOAA Atmospheric Models to Improve Ionospheric and Tropospheric Corrections," *U.S. Coast Guard DGPS Symp.*, Portsmouth, VA, June 19, 2003.

[25] Holub,K., "GPS Meteorology in Alaska," http://www.gpsmet.noaa.gov/jsp/downloads/GPS-Met_Anchorage.ppt, 2005.

[26] Kouba, J., "A Guide to Using International GPS Service (IGS)," 2003, http://igscb.jpl.nasa.gov/igscb/resource/pubs/GuidetoUsingIGSProducts.pdf.

[27] Héroux, P. et al., "Products and Applications for Precise Point Positioning—Moving Towards Real-Time," *Proceedings of the ION GNSS 17th International Technical Meeting of the Satellite Division*, Long Beach, CA, September 21–24, 2004.

[28] Langley, R. B., "Dilution of Precision," *GPS World*, Vol. 10, No. 5, May 1999, pp. 52–59.

5

GPS Positioning Modes

Positioning with GPS can be performed in either of two ways: point (absolute) positioning or relative positioning. Classical GPS point positioning employs one GPS receiver that uses the measured code pseudoranges and the broadcast ephemeris to determine the user's position instantaneously, as long as four or more satellites are visible at the receiver. The expected horizontal positioning accuracy from the civilian C/A-code receivers has increased (improved) from about 100m (2 drms) when SA was on to about 22m (2 drms) in the absence of SA [1]. As stated in Chapter 2, an additional 15 percent to 20 percent improvement in the positioning accuracy is expected as a result of the recent upgrade in the ground control segment. Classical GPS point positioning is used mainly when a relatively low accuracy is required. This includes recreation applications and low-accuracy navigation. Recently, PPP was introduced, which uses ionosphere-free linear combinations of carrier-phase and pseudorange measurements along with precise ephemeris and clock products. PPP provides positioning accuracy comparable to that of relative positioning (i.e., centimeter to decimeter accuracy).

GPS relative positioning, on the other hand, employs two GPS receivers simultaneously tracking the same satellites. If both receivers track at least four common satellites, a positioning accuracy level on the order of a few meters to millimeters can be obtained [2]. Carrier-phase and pseudorange measurements can be used in GPS relative positioning, depending on the accuracy requirements. The former provides the highest possible accuracy. GPS relative positioning can be made in either real-time or post-mission modes. GPS relative positioning is used for high-accuracy applications such as surveying and mapping, GIS, and precise navigation.

5.1　GPS Point Positioning: The Classical Approach

GPS point positioning, also known as standalone or autonomous positioning, involves only one GPS receiver. That is, one GPS receiver simultaneously tracks four or more GPS satellites to determine its own coordinates with respect to the center of the Earth (Figure 5.1). Almost all of the GPS receivers currently available on the market are capable of physically displaying their point-positioning coordinates.

To determine the receiver's point position at any time, the satellite coordinates as well as a minimum of four ranges to four satellites are required [2]. The receiver gets the satellite coordinates through the navigation message, while the ranges are obtained from either the C/A-code or the P(Y)-code, depending on the receiver type (civilian or military). As mentioned before, the measured pseudoranges are contaminated by both the satellite and receiver clock synchronization errors with respect to the stable GPS time. Correcting the satellite clock errors may be done by applying the satellite clock correction in the navigation message; the receiver clock error is treated as an additional unknown parameter in the estimation process [2]. This brings the total number of unknown parameters to four: three for the receiver coordinates and one for the receiver clock error. This is the reason

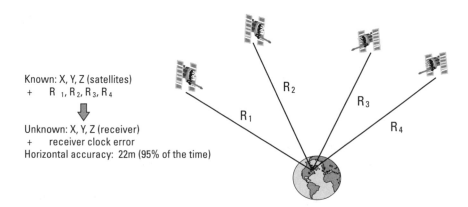

Figure 5.1　Principle of GPS point positioning.

that at least four satellites are needed. It should be pointed out that if more than four satellites are tracked, the so-called least-squares estimation or Kalman filtering technique is applied, which enables all measurements to contribute to the position solution [2–4]. As the satellite coordinates are given in the WGS 84 system, the obtained receiver coordinates will be in the WGS 84 system as well, as explained in Appendix A. However, most GPS receivers provide the transformation parameters between WGS 84 and many local datums used around the world.

To demonstrate the performance of the classical point positioning approach, GPS data collected at Algonquin, a continuously tracking site of the Canadian Active Control System (CACS) network, was processed in the point positioning mode. The data was collected on November 1, 2002, spanned 24 hours, and was processed using the GPSPace software, which was developed by NRCan. Figure 5.2 shows the residual errors (the difference between the approximate, or known, and the estimated values) in the latitude, longitude, and height components, when the L1 code pseudorange and broadcast ephemeris were used. Note that Figure 5.2 represents the epoch-by-epoch solution, which means that only the measurements of a particular epoch are used to generate a positioning solution for that epoch. It can be seen that the error in either of the horizontal components (i.e., latitude and longitude) reaches a maximum of 10m, while the error in the height component reaches up to about 19m. As explained in Chapter 4, the GPS height solution is expected to be less precise than the horizontal solution because of the satellite geometry effect.

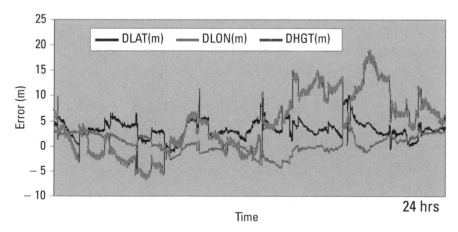

Figure 5.2 Epoch-by-epoch results (L1 code pseudorange and broadcast ephemeris).

5.2 GPS PPP

The accuracy of classical GPS point positioning is limited as a result of the presence of unmodeled errors and biases (see Chapter 4). In the absence of SA, ionospheric delay becomes the largest contributor to the GPS error budget [1]. As shown in Chapter 2, using a dual-frequency receiver, we can combine the L1 and L2 measurements to generate the ionosphere-free linear combination, which effectively removes the ionospheric delay. Figure 5.3 shows the residual errors in the latitude, longitude, and height components for the same data set described in Section 5.1, when the ionosphere-free undifferenced pseudorange and broadcast ephemeris were used. By comparing the obtained results with the classical point positioning results, it can be seen that the solution has improved in all three components. Further improvement to the point positioning solution could be attained through the use of precise satellite ephemeris and clock data produced by, for example, the IGS (Figure 5.4). In fact, until recently such an accuracy level was only achievable through code-based differential GPS (see Section 5.8).

To achieve the highest possible point positioning accuracy, both carrier-phase and pseudorange measurements should be used. In addition, the remaining unmodeled errors, including receiver clock error, tropospheric delay, satellite antenna offsets, phase wind-up, site displacement effects, and equipment delays, must be dealt with. This approach, which may utilize undifferenced or between-satellite single difference measurements, is commonly known as PPP [5–7]. Fortunately, most of these remaining errors can

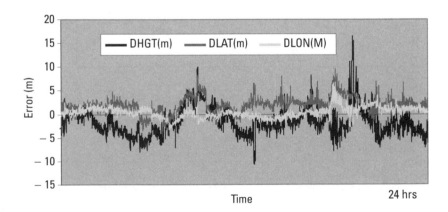

Figure 5.3 Results for ionosphere-free, undifferenced pseudorange with broadcast ephemeris.

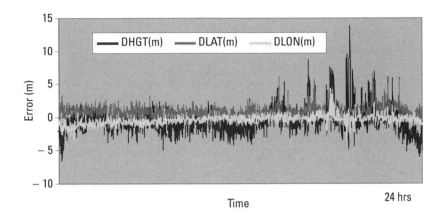

Figure 5.4 Results for ionosphere-free, undifferenced pseudorange with precise ephemeris and clocks.

be modeled with sufficient accuracy [8]. Exceptions are the receiver clock error (in the case of undifferenced measurements) and tropospheric delay, which are usually treated as additional unknowns to be estimated along with the station coordinates and ambiguity parameters. The same data set described previously was processed again using the GPS PPP approach. However, unlike the earlier cases, a sequential least-squares filter was used in this case. This means that as soon as new GPS measurements are collected at a particular epoch, they are used to improve the positioning solution in a sequential manner. Figure 5.5 shows the residual errors in the latitude, longitude, and height components. It can be seen that the PPP solution converges after some number of epochs to approach, within several centimeters, the precisely known station coordinates. It should be emphasized, however, that the PPP solution was based on a sequential processing (i.e., not epoch by epoch) and that the receiver was known to be stationary (i.e., static). As shown by [9], ignoring the receiver dynamics (e.g., a receiver onboard a vessel) would certainly degrade the accuracy.

Unfortunately, currently there is a delay in getting access to precise orbit and clock products, though a number of organizations (e.g., NRCan and the JPL) have started to make the orbit and clock corrections available in real time over the Internet. In addition, not all errors and biases are mitigated in the ionosphere-free linear combinations. This leaves residual error components unmodeled, which slows down the convergence of the PPP solution. At present, a PPP solution takes minutes to hours to converge to an accuracy at the decimeter to centimeter level, with a kinematic solution taking at least

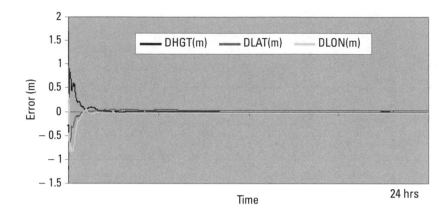

Figure 5.5 Results for GPS PPP.

twice as long as that of a static solution [5]. Clearly, this is not suitable for real-time applications. Nevertheless, the technique is promising, and research is underway to overcome these drawbacks.

5.3   GPS Relative Positioning

GPS relative positioning, also called differential positioning, employs two (or more) GPS receivers simultaneously tracking the same satellites to determine their relative coordinates (Figure 5.6). Of the two receivers, one is selected as a reference, or base, which remains stationary at a site with precisely known coordinates (i.e., presurveyed). The coordinates of the other receiver, known as the rover or remote receiver, are unknown. They are determined relative to the reference using measurements recorded simultaneously at the two receivers. The rover receiver may or may not be stationary, depending on the type of the GPS operation.

   A minimum of four common satellites is required for relative positioning. However, tracking more than four common satellites simultaneously would improve the precision of the GPS position solution [2]. Carrier-phase and/or pseudorange measurements can be used in relative positioning. A variety of positioning techniques are used to provide positioning information in real time or at a later time (i.e. postprocessing). Details of the commonly used relative positioning techniques are given in Sections 5.4 through 5.8. Generally, GPS relative positioning provides a higher accuracy than that of autonomous positioning. Depending on whether the pseudorange or

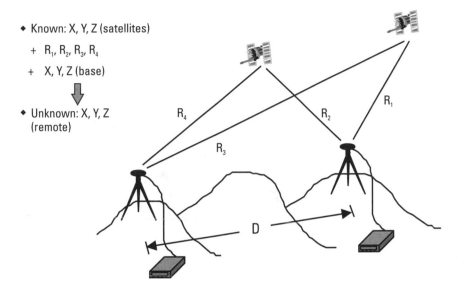

- Known: X, Y, Z (satellites)
  + R_1, R_2, R_3, R_4
  + X, Y, Z (base)

- Unknown: X, Y, Z (remote)

R_4 R_3 R_2 R_1 D

Figure 5.6 Principle of GPS relative positioning.

carrier-phase measurements are used in relative positioning, an accuracy level of a few meters to millimeters, respectively, can be obtained. This is mainly because the measurements of two (or more) receivers simultaneously tracking a particular satellite contain more or less the same errors and biases [10]. The shorter the distance between the two receivers is, the more similar the errors. Therefore, if we take the difference between the measurements of the two receivers (hence the name *differential positioning*), common errors will be removed and those that are spatially correlated will be reduced depending on the distance between the reference receiver and the rover (see Section 2.9).

5.4 Static GPS Surveying

Static GPS surveying is a relative positioning technique that depends primarily on the carrier-phase measurements [2]. It employs two (or more) stationary receivers simultaneously tracking the same satellites (see Figure 5.7). One receiver, the base receiver, is set up over a point with precisely known coordinates such as a survey monument (sometimes referred to as the known point). The other receiver, the remote receiver, is set up over a point whose coordinates are sought (sometimes referred to as the unknown point). The base receiver can support any number of remote receivers, as long as a

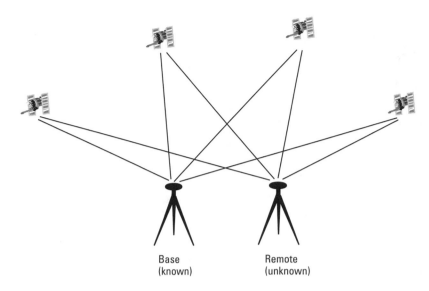

Base   Remote
(known)   (unknown)

Figure 5.7   Static GPS surveying.

minimum of four common satellites is visible at both the base and the remote sites.

In principle, this method is based on collecting simultaneous measurements at both the base and remote receivers for a certain period of time, which, after processing, yield the coordinates of the unknown point. The observation, or occupation time, varies from about 20 minutes to a few hours or more, depending on the distance between the base and the remote receivers (i.e., the baseline length), the number of visible satellites, and the satellite geometry. The measurements are usually taken at a recording interval of 15 seconds or more (i.e., one sample measurement every 15 seconds or more).

After completing the field measurements, the collected data is downloaded from the receivers onto a PC for processing. Different processing options may be selected depending on the user requirements, the baseline length, and other factors. For example, if the baseline is relatively short, say, 15 or 20 km, resolving the ambiguity parameters would be a key issue to ensure high-precision positioning. As such, in this case, the option of fixing the ambiguity parameters (i.e., determination of the initial integer ambiguity parameters; see Chapter 6) should be selected. In contrast, if the baseline is relatively long, a user may select the ionosphere-free linear combination option to remove the majority of the ionospheric error (see Chapter 2 for details on the various linear combinations of the GPS observables). This is because the ambiguity parameters may not be fixed reliably at the correct integer values, as a result of remaining unmodeled errors and biases. For long baselines (e.g., several hundreds of kilometers or more), it is recommended that the user processes the data with one of the scientific software packages available, such as the BERENSE software developed by the University of Bern, rather than a commercial software package. The precise ephemeris should also be used in this case, as the effect of the orbital errors will be considerably different at the two ends of the baseline. In fact, the use of PPP may be more appropriate in some cases (e.g., finding the precise coordinates of a point in a very remote area with no nearby base stations).

Static GPS surveying with the carrier-phase measurements is the most accurate positioning technique. This is mainly due to the considerable change in satellite geometry over the long observation time span. Although both the single- and dual-frequency receivers can be used for static positioning, the latter is often used, especially for baselines exceeding 20 km. The expected accuracy from a geodetic-grade receiver is typically 5 mm + 0.5 ppm (rms) for the horizontal component and 5 mm + 1 ppm (rms) for the vertical component, with ppm standing for parts per million and rms for root-mean-square. That is, for a 10-km baseline, for example, the expected accuracy of the static GPS surveying is 1.0 and 1.5 cm (rms) for the horizontal and vertical components, respectively. Higher accuracy may be obtained by, for example, applying the precise ephemeris.

5.5 Fast (Rapid) Static

Fast, or rapid, static surveying is a carrier phase–based relative positioning technique similar to static GPS surveying. That is, it employs two or more receivers simultaneously tracking the same satellites. However, with rapid static surveying, only the base receiver remains stationary over the known point during the entire observation session (see Figure 5.8). The rover receiver remains stationary over the unknown point for a short period of time only and then moves to another point whose coordinates are sought [2]. Similar to the static GPS surveying, the base receiver can support any number of rovers.

This method is suitable when the survey involves a number of unknown points located in the vicinity (i.e., within up to about 20 km) of a known point. The survey starts by setting up the base receiver over the known point, while setting up the rover receiver over the first unknown point (Figure 5.8). The base receiver remains stationary and collects data continuously. The rover receiver collects data for a period of about 2 to 10

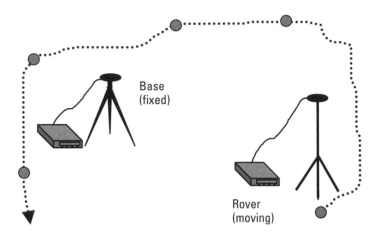

Figure 5.8 Fast (rapid) static GPS surveying.

minutes, depending on the distance to the base as well as the satellite geometry [2]. Once the rover receiver has collected the data, the user moves to the following point with unknown coordinates and repeats the procedures. It should be pointed out that, while moving, the rover receiver may be turned off. Due to the relatively short occupation time for the rover receiver, the recording interval is reduced to 5 seconds or less.

After collecting and downloading the field data from both receivers, the PC software is used for data processing. Depending on whether enough common data was collected, the software may output a fixed solution, which indicates that the ambiguity parameters were fixed at integer values (see Chapter 6 for details). Otherwise, a float solution is obtained, which means that the software was unable to fix ambiguity parameters at integer values (i.e., only real-valued ambiguity parameters were obtained). This problem occurs mainly when the collected GPS data is insufficient. A fixed solution means that the positioning accuracy is at the centimeter level, while the float solution means that the positioning accuracy is likely to be at the decimeter or submeter level. Although both the single- and dual-frequency receivers can be used for rapid static surveying, the probability of getting a fixed solution is higher with the latter.

5.6 Stop-and-Go GPS Surveying

Stop-and-go surveying is another carrier phase–based relative positioning technique. It also employs two or more GPS receivers simultaneously tracking the same satellites (Figure 5.9): a base receiver that remains stationary over the known point and one or more rover receivers [2]. The rover receiver travels between the unknown points and makes a brief stop at each point to collect the GPS data. The data is usually collected at a 1- to 2-second recording rate for a period of about 30 seconds per each stop. Similar to the previous methods, the base receiver can support any number of rovers. This method is suitable when the survey involves a large number of unknown points located in the vicinity (i.e., within up to 15–20 km) of a known point.

The survey starts by first determining the initial integer ambiguity parameters, a process known as receiver initialization. This could be done by various methods, discussed in the next chapter. Once the initialization is performed successfully, centimeter-level positioning accuracy can be obtained instantaneously. This is true as long as there is a minimum of four common satellites simultaneously tracked by both the base and the rover receivers at all times. If this condition is not fulfilled at any moment during the survey, the initialization process must be repeated to ensure centimeter-level accuracy.

Following the initialization, the rover moves to the first unknown point. After collecting about 30 seconds of data, the rover moves, without being switched off, to the second point, and the procedure is repeated. It is of the utmost importance that at least four satellites are tracked, even during the move; otherwise, the initialization process must be repeated by, for example, reoccupying the previous point. Some manufacturers recommend the reoccupation of the first point at the end of the survey. This is very useful in obtaining a fixed solution, or centimeter-level accuracy, provided that the

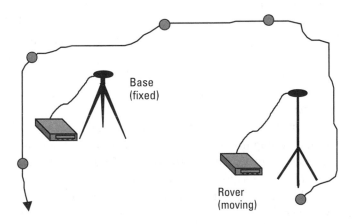

Figure 5.9 Stop-and-go GPS surveying.

processing software has forward and backward processing functions. Once the data is collected and downloaded, the PC software is used to process it. Both single- and dual-frequency receivers may use the stop-and-go surveying method.

A special case of stop-and-go surveying is known as kinematic GPS surveying, also known as conventional or postprocessed kinematic (PPK) GPS. Both methods are the same in principle; however, the latter requires no stops at the unknown points. The positional accuracy is expected to be higher with the stop-and-go surveying, as the errors are averaged out when the receiver stops at the unknown points.

5.7 RTK GPS

RTK GPS surveying is a carrier phase–based relative positioning technique that, like the previous methods, employs two (or more) receivers simultaneously tracking the same satellites (Figure 5.10). This method is suitable when (1) the survey involves a large number of unknown points located in the vicinity (i.e., within up to about 15–20 km) of a known point; (2) the coordinates of the unknown points are required in real time; and (3) the line of sight, the propagation path, between the two receivers is relatively unobstructed [11]. Because of its ease of use as well as its capability to determine the coordinates in real time, this method is preferred by many users.

In this method, the base receiver remains stationary over the known point and is commonly attached to a radio transmitter (Figure 5.10). The rover receiver is normally carried in a backpack (or mounted on a moving object) and is attached to a radio receiver. Similar to the conventional kinematic GPS method, a data rate of 1 Hz (one sample per second) or lower is normally employed. The base receiver measurements and coordinates (along with the antenna height) are transmitted to the rover receiver through the communication (radio) link [12, 13]. The built-in software in the rover receiver combines and processes the GPS measurements collected at both the base and the rover receivers to obtain the rover coordinates.

The initial ambiguity parameters are determined almost instantaneously using a technique called on-the-fly (OTF) ambiguity resolution, to be discussed in the next chapter. Once the ambiguity parameters are fixed to integer values, the receiver (or its handheld computer controller) will display the rover coordinates right in the field. That is, no postprocessing is required. The expected positioning accuracy is of the order of 10 mm + 1 ppm (rms) for the horizontal component and 20 mm + 1 ppm (rms) for the vertical

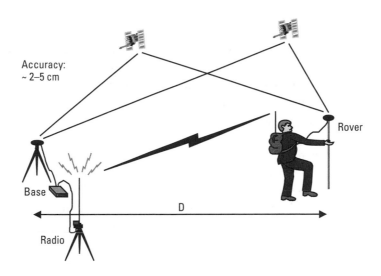

Figure 5.10 RTK GPS surveying.

component. This can be improved by staying over the point for a short period of time (e.g., about 30 seconds) to allow for averaging the position. The computed rover coordinates for the entire survey may be stored and downloaded at a later time into computer-aided design (CAD) software for further analysis. This method is used mainly, but not exclusively, with dual-frequency receivers.

Under the same conditions, the positioning accuracy of the RTK method is slightly degraded compared with that of the conventional kinematic GPS method. This is mainly because the time tags (or time stamps) of the conventional kinematic data from both the base and the rover match perfectly in the processing. With RTK, however, the base receiver data reaches the rover after some delay (or latency). Data latency occurs as a result of formatting, packetizing, transmitting, and decoding the base data [12]. To match the time tag of the rover data, the base data must be extrapolated, which degrades the positioning accuracy.

5.8 Real-Time Differential GPS

Real-time differential GPS (DGPS) is a code-based relative positioning technique that employs two or more receivers simultaneously tracking the same satellites (Figure 5.11). It is used when real-time meter-level accuracy is required. The method is based on the fact that some of the main GPS errors in the measured pseudoranges are essentially the same at both the base and the rover, as long as the baseline length is within a few hundred kilometers.

As before, the base receiver remains stationary over the known point. The built-in software in the base receiver uses the precisely known base coordinates as well as the satellite coordinates, derived from the navigation message, to compute the ranges to each satellite in view. The software further takes the difference between the computed ranges and the measured code pseudoranges to obtain the pseudorange errors (or DGPS corrections). These corrections are transmitted in a standard format called Radio Technical Commission for Maritime Service (RTCM) to the rover through a communication link (see Chapter 8 for more about RTCM). The rover then applies the DGPS corrections to correct the measured pseudoranges at the rover. Finally, the corrected pseudoranges are used to compute the rover coordinates.

The accuracy obtained with this method varies between a submeter and about 5m, depending on the base-rover distance, the transmission rate of the RTCM DGPS corrections, and the performance of the C/A-code

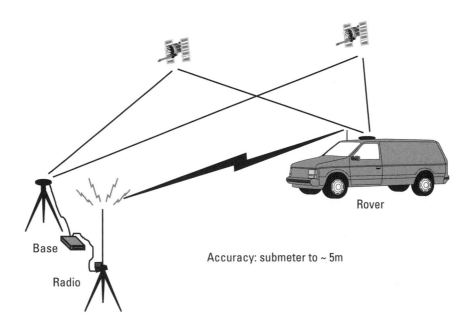

Figure 5.11 Real-time DGPS operation.

receivers [2]. Higher accuracy is obtained with short base-rover separation, high transmission rate, and carrier-smoothed C/A-code ranges. With the termination of SA, the data rate could be reduced to 10 seconds or less without noticeable accuracy degradation. Further accuracy improvement could be achieved if the receivers are capable of storing the raw pseudorange measurements, which could be used at a later time in the postprocessing mode. As the real-time DGPS is widely used, some governmental agencies as well as private firms provide the RTCM DGPS corrections either at no cost or for a fee. Further information on these services will be given in Chapter 7.

5.9   Real Time Versus Postprocessing

The term *real time* means that the results are obtained almost instantaneously, while the term postprocessing means that the measurements are collected in the field and processed at a later time to obtain the results. Each of these modes has some advantages and some disadvantages.

The first advantage of the real-time mode is that the results as well as the accuracy measures (or quality control) are obtained while in the field. This is especially important for RTK surveying, as the user would not store the displayed coordinates unless the ambiguity parameters are shown to be fixed at integer values and centimeter-level accuracy is achieved. This leads to a higher productivity compared with the postprocessing mode, as only enough GPS data to obtain a fixed solution is collected. In addition, processing the GPS data is done automatically in the field by the built-in software. This means that no postprocessing software training is required. The user also saves the time spent in data processing.

There are, however, some advantages in the postprocessing mode as well. The first of these is that more accurate results are generally obtained with the postprocessing mode. One reason for this is more flexibility in editing and cleaning of the collected GPS data. As well, there is no accuracy degradation due to data latency, as explained in Section 5.8. Another important advantage is that communication link–related problems, such as signal obstruction or limitation of coverage, are avoided. In some cases, the input parameters, such as the base station coordinates or the antenna height, may contain some errors, which lead to errors in the computed rover coordinates. These errors can be corrected in the postprocessing mode, while they cannot be completely corrected in the real-time mode.

5.10   Communication (Radio) Link

RTK and real-time DGPS operations require a communication, or radio, link to transmit the information from the base receiver to the rover receiver (Figures 5.10 and 5.11). Both ground- and satellite-based communication links are used for this purpose. However, we will restrict our discussion in this section to dedicated ground-based communication links, which are widely used. Information about satellite-based communication links are given in Chapter 7. RTK data are typically transmitted at a baud rate of 9,600, while the DGPS corrections are typically transmitted at 200 Kbps. A variety of radio links that use different parts of the electromagnetic spectrum are available to support such operations. The spectrum parts mostly used in practice are the low/medium frequency (LF/MF) bands (i.e., 30 kHz to 3 MHz) and the very high and ultrahigh frequency (VHF/UHF) bands (i.e., 30 MHz to 3 GHz) [12, 13]. Often, GPS users utilize their own dedicated radio links to transmit base station information.

Dedicated ground-based GPS radio links are mostly established using the VHF/UHF band. Radio links in this band provide line-of-sight coverage, with the ability to penetrate into buildings and other obstructions. One example of such a radio link is the widely used Position Data Link (PDL) from Pacific Crest Corporation (see Figure 5.12). PDL allows for a baud rate of 19,200 and is characterized by low power consumption and enhanced user interface. This type of radio link requires a license to operate. Another example is the license-free spread-spectrum radio transceiver, which operates in the 902–928 MHz portion of the UHF band (Figure 5.12). This radio link has coverage of 1–5 km and 3–15 km in urban and rural areas, respectively. More recently, some GPS manufacturers adopted cellular technology, the

PDL radio   Spread spectrum radio

Figure 5.12 Examples of radio modems. (*Courtesy of:* Magellan Corporation.)

digital personal communication services (PCS) and the third generation (3G) wideband digital networks, as an alternative communication link. The 3G technology uses common global standards, which reduces the service cost. In addition, this technology allows the devices to be kept in the "on" position all the time for data transmission or reception, while the subscribers pay for the packets of data they transmit/receive. A number of GPS receivers currently available on the market have built-in radio modems that use one or more of these technologies.

It should be pointed out that obstructions along the propagation path, such as buildings and terrain, attenuate the transmitted signal, which leads to limited signal coverage. The transmitted signal attenuation may also be caused by ground reflection (multipath), the transmitting antenna, and other factors [12]. To increase the coverage of a radio link, a user may employ a power amplifier or high-quality coaxial cables, or he or she may increase the height of the transmitting/receiving radio antenna. If a user employs a power amplifier, however, he or she should be cautioned against signal overload, which usually occurs when the transmitting and the receiving radios are very close to each other [12].

A user may also increase the signal coverage by using a repeater station. In this case, it might be better to use a unidirectional antenna, such as a Yagi, at the base station and an omnidirectional antenna at the repeater station (see Figure 5.13) [13].

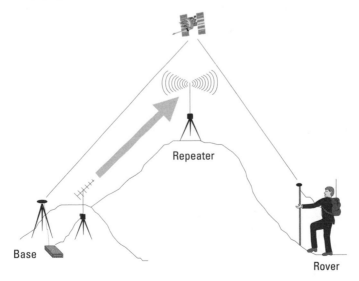

Figure 5.13 Use of repeaters to increase the radio coverage.

References

[1] Shaw, M., K. Sandhoo, and D. Turner, "Modernization of the Global Positioning System," *GPS World*, Vol. 11, No. 9, September 2000, pp. 36–44.

[2] Hoffmann-Wellenhof, B., H. Lichtenegger, and J. Collins, *Global Positioning System: Theory and Practice*, 3rd ed., New York: Springer-Verlag, 1994.

[3] Leick, A., *GPS Satellite Surveying. 3rd edition.* New Jersey: John Wiley & Sons, 2004.

[4] Levy, L. J., "The Kalman Filter: Navigation's Integration Workhorse," *GPS World*, Vol. 8, No. 9, September 1997, pp. 65–71.

[5] Héroux, P. et al., "Products and Applications for Precise Point Positioning—Moving Towards Real-Time," *Proceedings of the ION GNSS 2004*, Long Beach, CA, September 21–24, 2004.

[6] Han, S.-C., J. H. Kwon, and C. Jekeli, "Accurate Absolute GPS Positioning Through Satellite Clock Error Estimation." *Journal of Geodesy*, Vol. 75, No. 1, 2001, pp. 33–43.

[7] Colombo, O. L., and A. W. Sutter, "Evaluation of Precise, Kinematic GPS Point Positioning," *Proceedings of the ION GNSS 2004*, Long Beach, CA, September 21–24, 2004, CD-ROM.

[8] Kouba, J., "A Guide to Using International GPS Service (IGS)," 2003, http://igscb.jpl.nasa.gov/igscb/resource/pubs/GuidetoUsingIGSProducts.pdf.

[9] Abdel-Salam, M., Y. Gao, and X. Shen, "Analyzing the Performance Characteristics of a Precise Point Positioning System," *Proceedings of the ION GPS 2002*, Portland, OR, September 24–27, 2002, CD-ROM.

[10] Langley, R. B., "The GPS Observables," *GPS World*, Vol. 4, No. 4, April 1993, pp. 52–59.

[11] Langley, R. B., "RTK GPS," *GPS World*, Vol. 9, No. 9, September 1998, pp. 70–76.

[12] Langley, R. B., "Communication Links for DGPS," *GPS World*, Vol. 4, No. 5, May 1993, pp. 47–51.

[13] Pacific Crest Corporation, "The Guide to Wireless GPS Data Links," 2000.

6

Ambiguity Resolution Techniques

The previous chapter showed that centimeter-level positioning accuracy could be achieved with the carrier-phase observables. A prerequisite to this, however, is the successful determination of the initial integer cycle ambiguity parameters (integer double-difference ambiguity parameters in the relative positioning mode). This process is commonly known as ambiguity resolution or ambiguity fixing. As discussed in Section 2.6, resolving the ambiguity parameters correctly is equivalent to having very precise ranges to the satellites, which leads to high-accuracy positioning. This chapter focuses on the ambiguity resolution for GPS relative positioning.

The ambiguity parameters are initially determined as part of a least-squares, or Kalman filtering, solution [1, 2]. Unfortunately, neither method can directly determine the integer numbers of the ambiguity parameters. This is mainly because the ambiguity parameters are determined from noisy data. What can be obtained are the real-valued numbers along with their uncertainty parameters (so-called covariance matrix) only. These real-valued numbers are in fact difficult to separate from the baseline (i.e., reference-rover separation) solution [3]. As such, since we know in advance that the ambiguity parameters are integer numbers, it becomes clear that further analysis is required.

Traditionally, high-precision GPS relative positioning with carrier-phase observables is carried out using long observational time spans (typically a few hours). This allows for the receiver-satellite geometry to change sufficiently to facilitate the separation of the ambiguity parameters from the baseline solution. As such, even though the least-squares solution contains real-valued numbers for the ambiguity parameters (sometimes called float solution), they are very close to integer values. Consequently, the correct integer values are simply obtained by rounding off the real-valued numbers to the nearest integers [4]. Another least-squares estimation process is then carried out, considering the integer-valued ambiguity parameters as known values (i.e., fixed) while the baseline components are unknowns. It is clear that, although this method is capable of determining the correct integer values of the ambiguity parameters, it is time consuming. As such, the use of this method is currently limited to long baselines in the static mode.

Various methods have been developed to overcome the limitation of the previous method (i.e., the use of long observational time spans). One such method is to use a known baseline (i.e., the coordinates of its end points are accurately known), which might be available within the project area. The ambiguity parameters are determined by simply occupying the two end points of the known baseline with the base and the rover receivers for a short period of time. This process is commonly known as receiver initialization. Following receiver initialization, the rover receiver can move to the points to be surveyed. With this method, the receiver uses the ambiguity parameters determined during the initialization to solve for the coordinates of the new points. As mentioned in Chapter 2, the initial integer number of cycles (the ambiguity parameter) remains constant over time, even if the receiver is in motion, as long as no cycle slips have occurred. In other words, it is necessary that the receivers be kept "on" all the time and that at least four common satellites are tracked at any moment. An alternative initialization method is known as the antenna swap method, which can be used when no known baseline is available within the project area. This method, which was introduced by Dr. Ben Remondi in 1986, is based on exchanging the antennas between the base and the rover while tracking at least four satellites. More details on this method are given later. Both the known baseline and the antenna swap methods are more suitable for kinematic positioning in the postprocessing mode.

The three methods discussed here are suitable for non-real-time applications where data are collected in the field and then processed at a later time (i.e. postprocessed). RTK positioning, however, requires that the integer ambiguity parameters be determined while the receiver is in motion, or OTF [5]. Resolving the ambiguities OTF, often called OTF ambiguity resolution, is different from the ambiguity-resolution techniques mentioned earlier in the sense that the initialization is performed in the field using observational time spans as short as a few seconds. Due to the high altitudes of the GPS satellites, the receiver-satellite geometry changes very slowly over time. As such, a short time span of data causes some difficulties in resolving the ambiguities. Fortunately, a more advanced technique has been developed to overcome this limitation; this technique is discussed later.

6.1   Antenna Swap Method

Antenna swap is a method used for a fast and reliable determination of the initial ambiguity parameters (i.e., initialization) in the postprocessing mode [6]. This method is used mainly when single-frequency receivers are used for kinematic surveying, although it can be used with dual-frequency receivers as well.

The initialization procedure with the antenna swap method starts with the setting up of the reference (base) receiver over the known point while setting up the rover within a few meters of it (Figure 6.1). About 1 minute of simultaneous GPS data are then collected by both receivers. Usually, a data rate as high as 1 or 2 seconds is used. Once the data are collected, the two antennas (with the two GPS receivers connected) are exchanged (see Figure 6.1). This is done without changing the original antenna heights. Care must be taken to continuously track a minimum of four, or preferably five, common satellites. With this new setup, the data capture is repeated in the same way as the previous step. After this, the receivers are returned to their original setup, which ends the initialization procedures. As can be observed, swapping the two antennas allows for sufficient geometrical changes, which is the key for resolving the ambiguity parameters.

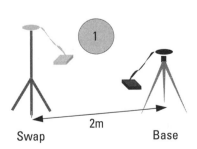 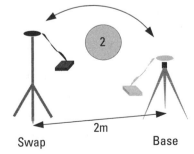

Figure 6.1 Antenna swap method.

Once the initialization is performed, the base receiver must be kept stationary over the known point while the rover moves between the points to be surveyed, as discussed earlier in the kinematic method. After finishing the fieldwork, the data is downloaded onto a computer running postprocessing software. This software first uses the initialization data to determine the initial ambiguity parameters. Once determined, the software will use these parameters to determine the coordinates of the survey points at centimeter-level accuracy. It should be pointed out that a shorter observational time span would be enough for the receiver initialization.

6.2   OTF Ambiguity Resolution

OTF ambiguity resolution is an advanced technique recently developed to determine the initial integer ambiguity parameters without static initialization (i.e., while the rover receiver is moving). This technique may be applied with either single- or dual-frequency data. However, resolving the ambiguities is faster and more reliable with dual-frequency data. It is used mainly for, but not restricted to, RTK operations.

Several OTF techniques have been developed since the early 1990s. Only one method is summarized here [4]. The base and rover measurements are combined in the double-differenced mode, and an initial estimation by, for example, the least-squares or Kalman filtering technique, is then performed. The outcome of this initial estimation is an initial rover position along with estimates (real values) for the ambiguity parameters and their uncertainty values, or the covariance matrix.

The covariance matrix can be represented geometrically to form a region, known as the confidence region, around the estimated real-valued ambiguity parameters [4]. The size of such a confidence region depends on the size of the uncertainty parameters of the ambiguities as well as the probability level used. The larger the uncertainty values and/or the probability level, the larger the size of the confidence region. The confidence region takes the shape of an ellipse if the number of the estimated parameters is two and an ellipsoid if it is three. If the number of estimated parameters is more than three, which is the case if the number of satellites is more than four, a confidence region of a hyperellipsoid is obtained.

Generally, a confidence region of a hyperellipsoid is formed around the estimated real-valued ambiguity parameters. Such a hyperellipsoid contains the likely integer ambiguity parameters at a certain probability level. For example, if a probability value of 99 percent is used to scale the hyperellipsoid, it means that there is a 99 percent chance that the true integer ambiguity parameters are located inside that hyperellipsoid. Since we know in advance that the ambiguity parameters must be integer numbers, we may draw (mathematically) gridlines that intersect at integer values inside the hyperellipsoid. If the grid spacing is selected to be equal to one carrier cycle, then the likely integer ambiguity parameters would be represented by one of the points of intersection inside the hyperellipsoid. Figure 6.2 simplifies this, using a two-dimensional case as an example. The hyperellipsoid is then used to search for the likely integer values for ambiguity parameters (i.e., all the points inside the confidence region with integer values). Based on statistical evaluation, only one point is selected as the most likely candidate for the integer ambiguity parameters.

Unfortunately, as shown by [7], the ambiguity parameters are highly correlated, which causes the hyperellipsoid to be very elongated. This, in turn, makes the performance of the integer ambiguity search rather poor. To improve the performance of the integer ambiguity search, the least-squares ambiguity decorrelation adjustment (LAMBDA) was developed by Professor P. J. G. Teunissen in 1993 [8]. The LAMBDA method transforms the hyperellipsoid search space into a hyperspherelike search space (i.e., it attempts to decorrelate the ambiguity parameters), which improves the performance of the integer ambiguity search. As a result of its efficiency and high success rate in identifying the correct ambiguity parameters, the LAMBDA method is widely used across the world [9]. Once the ambiguity parameters are correctly resolved, they are introduced as known values in a

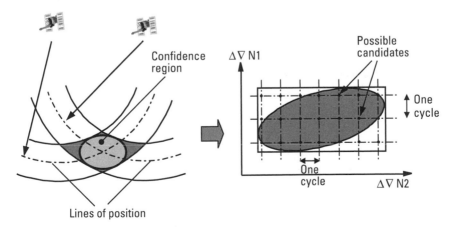

Figure 6.2 OTF ambiguity resolution.

final round of least-squares (or Kalman filtering) estimation, which is performed to obtain the rover coordinates at centimeter-level accuracy. It should be pointed out that the OTF technique, although designed mainly for resolving the ambiguity parameters in real time, could also be used in the non-real-time mode.

References

[1]   Leick, A., *GPS Satellite Surveying*, 3rd Edition. New Jersey: John Wiley & Sons, Inc., 2004.

[2]   Levy, L. J., "The Kalman Filter: Navigation's Integration Workhorse," *GPS World*, Vol. 8, No. 9, September 1997, pp. 65–71.

[3]   Teunissen, P. J. G., P. J. de Jonge, and C. J. M. Tiberius, "A New Way to Fix Carrier-Phase Ambiguities," *GPS World*, Vol. 6, No. 4, April 1995, pp. 58–61.

[4]   El-Rabbany, A., *The Effect of Physical Correlations on the Ambiguity Resolution and Accuracy Estimation in GPS Differential Positioning*, Technical Report No. 170, Department of Geodesy and Geomatics Engineering, Fredericton, New Brunswick, Canada: University of New Brunswick, 1994.

[5]   DeLoach, S. R., D. Wells, and D. Dodd, "Why On-the-Fly?" *GPS World*, Vol. 6, No. 5, May 1995, pp. 53–58.

[6]   Remondi, B., "Performing Centimeter-Level Surveys in Seconds with GPS Carrier Phase: Initial Results," *Proc. 4th Intl. Geodetic Symposium on Satellite Positioning*, Vol. 2, Austin, TX, April 28–May 2, 1986, pp. 1229–1249.

[7]   Teunissen, P. J. G., and A. Kleusberg (Eds.), *GPS for Geodesy*, 2nd Edition, New York: Springer-Verlag, 1998.

[8]   Hoffmann-Wellenhof, B., H. Lichtenegger, and J. Collins, *GPS Theory and Practice*, 5th rev. ed., New York: Springer-Verlag, 2001.

[9]   Joosten, P., and C. Tiberius, "LAMBDA: FAQ," *GPS Solutions*, Springer-Verlag, Vol. 6, 2002, pp. 109–5114.

7

GPS Data, Products, and Correction Services

GPS users employ differential techniques to achieve the highest possible positional accuracy. A straightforward manner of doing this is to use two GPS receivers, a base and a rover, as discussed in Chapter 5. However, this may not be cost effective in many instances. An alternative, which could significantly reduce the cost of a survey without degrading the positional accuracy, is to use one of the available GPS data and correction services. If such a service is available within the project area, a GPS user would only need one receiver to be used as a rover; no base receiver is required. A number of various GPS services are currently available with various levels of accuracy and cost. Some services are even provided with no user fees required.

For high-accuracy positioning (e.g., establishing new control points), GPS users may use one of the highly precise permanent GPS reference station networks established by several organizations around the world. These GPS data services are currently available free of charge at the global level, such as the IGS network, as well as at the regional level, such as the Continuously Operating Reference Station (CORS) network in the United States [1–3]. The CACS and the European Reference Frame (EUREF) Permanent Network (EPN) are other regional GPS data services, which are available to users at no cost [4, 5]. The reference stations within these systems operate on a continuous basis and provide access to the modern reference frames, such as the International Terrestrial Reference Frame (ITRF) and the improved North American Datum (NAD) 83 (see Appendix A for information about these systems). In addition to GPS data service, some organizations make available other very useful products, such as precise GPS orbits, precise satellite clock information, Earth rotation parameters, and ionospheric and tropospheric corrections.

Some countries around the world have established networks of reference stations around their coastal areas, which continuously broadcast real-time DGPS corrections in a special format known as the RTCM format (see Chapter 8). Each reference station operates independently of the other stations in the network to serve users within its coverage area, which makes it a single station–based DGPS service. This service is primarily designed to enhance the safety of marine navigation, but it is available at no cost to all users within the coverage area. Although this service is available largely free of charge, it requires a beacon receiver connected to a GPS receiver that accepts the RTCM corrections [6]. GPS receivers that accept the RTCM corrections are commonly known as differential-ready GPS receivers. The accuracy obtained from such a service is usually in the range of decimeters to a few meters.

Another real-time DGPS correction service is known as wide-area differential GPS (WADGPS). The system architecture of WADGPS involves a set of ground reference stations that cover a wide geographical area (e.g., a continent). Typically, the measurements from each reference station are processed at a central station to produce differential corrections valid over a large geographical area (i.e., within the system coverage area). The differential corrections are then transmitted to users through geostationary satellites. Similar to single-station DGPS systems, WADGPS systems require a special receiver to decode the DGPS correction information, which would be interfaced to the GPS rover receiver to output positional information at the meter-level accuracy. WADGPS systems have several advantages over conventional single-station DGPS systems, including coverage of large, inaccessible regions using fewer reference stations [7].

Multisite, real-time, carrier phase–based RTK positioning at subdecimeter-level accuracy is a relatively new service that has been recently developed [8]. With this service, as little as four GPS reference receivers could cover an entire city or even a number of adjacent small cities. Advancements in wireless communication and the Internet are expected to make this service very promising [9]. This chapter summarizes each of the GPS services.

7.1 GPS Data and Product Services

Several organizations around the world have established highly precise permanent GPS reference station networks (some with Russian satellite-based navigation system GLONASS as well), which are used for various geodetic purposes. The GPS data collected at these reference stations is made available to users and could be used for high-accuracy positioning operations, such as establishing new control points. One such organization is the IGS, which is a service with international multiagency membership to support global geodetic and geophysical activities [1]. The International Association of Geodesy (IAG) has formally recognized the IGS since 1993. The IGS is based on a global network of over 350 tracking stations (as of January 2006) equipped with continuously operating dual-frequency receivers and a number of operational, data, and analysis centers. In addition, the Central Bureau Information Systems (CBIS) consisting of a Web site and ftp server makes available some of the IGS products [10]. Figure 7.1 shows the current IGS tracking stations.

The raw GPS data collected at each IGS tracking station is formatted in a standard format, called the Receiver Independent Exchange (RINEX) (see details in Chapter 8), by the operational centers. The formatted data is then collected by the IGS data centers for archiving and online access (see [10] for more details). At this stage, the analysis centers use the online data to create a

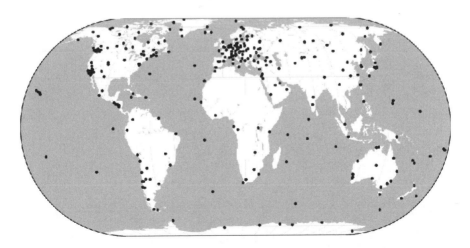

Figure 7.1 IGS tracking stations. (*Source:* http://igscb.jpl.nasa.gov/network/ netindex.html.)

number of products. These products include GPS and GLONASS precise ephemeris, satellite, and tracking station clock information; tracking station coordinates and velocities; Earth rotation parameters; global ionospheric maps; and zenith tropospheric path delay estimates. There are different accuracy levels for the IGS products depending on the time of availability, as explained in Chapter 4 [11]. Table 7.1 shows an example the current IGS products, namely GPS ephemeris and clock parameters.

IGS GPS data and products are available, at no cost, to users worldwide through the Internet (the IGS URL address is given in Appendix C). It should be pointed out that both the IGS precise ephemeris and the tracking station coordinates are referred to the ITRF reference system. That is, if a

Table 7.1
IGS GPS Ephemeris and Clock Products

GPS Satellite Ephemeris/ Satellite and Station Clocks	Product	Accuracy	Latency	Updates	Sample Interval
Broadcast	Orbits	~160 cm	Real time	–	Daily
	Satellite clocks	~7 ns			
Ultrarapid (predicted half)	Orbits	~10 cm	Real time	Four times daily	15 min
	Satellite clocks	~5 ns			
Ultrarapid (observed half)	Orbits	<5 cm	3 hours	Four times daily	15 min
	Satellite clocks	~0.2 ns			
Rapid	Orbits	<5 cm	17 hours	Daily	15 min
	Satellite and station clocks	0.1 ns			5 min
Final	Orbits	<5 cm	~13 days	Weekly	15 min
	Satellite and station clocks	<0.1 ns			5 min

Source: http://igscb.jpl.nasa.gov/components/prods.html.

user employs the IGS precise ephemeris, his or her solution coordinates will be referred to the ITRF reference system. In North America, users could also access GPS data through the CORS and CACS networks, which are operated by the U.S. National Geodetic Survey (NGS) and Natural Resources Canada, respectively [3, 4]. One advantage of using CORS and CACS data is that the reference stations are relatively closer to each other than the IGS stations. Additionally, the precise ephemeris and the coordinates of the reference stations may be obtained in either the ITRF or the improved NAD 83 system.

7.2 Maritime DGPS Service

Marine radio beacons are electronic aids to navigation that operate in the low-to-medium frequency band of 283.5–325 kHz [6]. They are installed at lighthouses and other coastal locations. To enhance maritime safety, a number of marine radio beacons throughout the world have been modified to broadcast real-time DGPS corrections in the RTCM format (see Figure 7.2). This service, which is considered as an augmentation to GPS, is available at no charge in most cases.

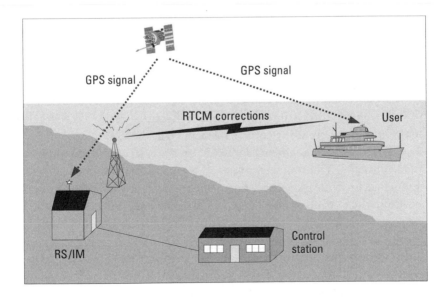

Figure 7.2 DGPS beacon service.

In the DGPS beacon system, a reference station creates the real-time DGPS corrections in the RTCM format, as discussed in Chapter 5. These corrections are digitally modulated using a special form of frequency modulation known as minimum shift keying (MSK). The modulated correction data is then transmitted from the radio beacon at rates between 25 and 200 bps [6]. Typical rates, however, are 100 and 200 bps. In most of the cases, an integrity-monitoring (IM) unit is co-located with the reference station to monitor its performance (see Figure 7.2).

A user equipped with an MSK beacon receiver can receive the transmitted DGPS corrections as long as he or she is within the coverage area of a particular beacon station. The coverage depends on, among other factors, the transmitter power output, the atmospheric noise, and the receiver sensitivity. The coverage also depends on the characteristics of the propagation path or conductivity; the coverage is greater over water than inland. Beacon locations are usually selected to provide overlapping coverage to increase the overall signal availability. The service provider (e.g., the coast guard of a particular country) usually publishes the expected coverage area of a DGPS beacon system. Some manufacturers of the radio beacon receivers publish detailed information about the availability of the beacon service worldwide.

It should be pointed out that MSK beacon receivers are available as single- or dual-channel receivers. The latter are more expensive but more reliable. The second channel is useful when searching for an adjacent beacon station, if available, with a better signal quality. The receiver will automatically switch to the adjacent beacon station once it is found. Evidently, if an area is known to be covered by one beacon station only, a single-channel MSK receiver would be sufficient and more cost-effective. To be useful, the MSK receiver should be interfaced to a differential-ready GPS receiver, which would then output the corrected station coordinates.

A beacon-based augmentation system provides a typical positional accuracy of better than one meter at each broadcast station. However, such positional accuracy degrades at a rate of about 1m for each 150 km of distance from the broadcast site [12]. Further accuracy degradation can be expected as a result of, for example, poor-quality user equipment and significant levels of multipath.

It is worth mentioning that the U.S. maritime DGPS Service has been recently expanded to a nationwide DGPS (NDGPS), which can be used for land navigation. By 2005, the U.S. Coast Guard, in cooperation with other federal agencies, had started the modernization of the NDGPS beacon-based augmentation system [13]. In addition to improving the quality of service, the modernization program is studying the option of transmitting the carrier-phase and pseudorange measurements from the NDGPS reference stations. Such service enhancement should improve the positional accuracy to the subdecimeter level.

7.3 WADGPS Systems

Real-time DGPS with a single reference station has the disadvantage that the positioning accuracy tends to deteriorate as the user moves away from the reference station. That is, the highest positioning accuracy is limited to the relatively small area surrounding the reference station. This is mainly because of the effect of orbital, ionospheric, and tropospheric errors (Chapter 4). To overcome this problem, a system based on a number of widely separated reference stations, known as WADGPS, has been developed [7]. Typically, WADGPS involves a network of widely separated reference stations, one or more master stations, uplink stations, and communication satellites. WADGPS can be categorized into three groups according to the algorithm used, namely position domain, measurement domain, and state-space domain [14]. With the position-domain WADGPS, a user receives DGPS corrections from several reference stations, which are used independently to obtain a number of position fixes. The user's estimated position is then obtained based on a weighted average of the resulting position fixes. This system is in fact an extension of the single-station (conventional) DGPS and is not widely used in practice.

A relatively more efficient system is the measurement-domain WADGPS. With this system, the network of reference stations collect and preprocess the GPS data to compute the DGPS corrections, which are then forwarded to a master station via, for example, terrestrial links such as fiber optic cables. The master station analyzes and combines the received data (using, for example, a weighted average or least-squares regression method) to determine a number of correction parameters for each GPS satellite, which would be valid within the system coverage area (see Figure 7.3). These parameters are packed and uploaded into a geostationary satellite, which rebroadcasts them back to the Earth to ensure a wide coverage. A user within the system coverage area will receive only one set of DGPS corrections, which is valid for his or her location (i.e., as if the user is in the close vicinity of a virtual reference station). This approach is relatively simple; however, the solution degrades slowly as the user moves away from the center of the network.

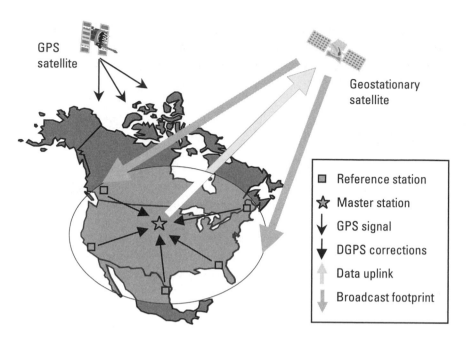

GPS satellite

Geostationary satellite

- Reference station
- Master station
- GPS signal
- DGPS corrections
- Data uplink
- Broadcast footprint

Figure 7.3 Principle of WADGPS system.

One example of the measurement-domain WADGPS is the Fugro's OmniSTAR system virtual base station (VBS), a code-based commercially available differential GPS correction service. According to OmniSTAR's Web site, the system uses over 90 reference stations throughout the world [15]. The OmniSTAR service operates in the L-band of the frequency spectrum. To access the service, a subscriber needs an OmniSTAR-enabled single-frequency GPS receiver (to receive and decode the DGPS corrections), which must be set to the proper frequency of the local OmniSTAR's geostationary satellite (MSV satellite for North America). The expected horizontal positioning accuracy of the OmniSTAR VBS service is better than one meter at the 95 percent probability level [16]. It should be pointed out that OmniSTAR provides other higher accuracy services. These are discussed at the end of this section and in Section 7.4.

The third approach, state-space-domain WADGPS, attempts to estimate each of the major GPS errors (e.g., satellite clock and orbital errors), based on the GPS measurements collected at the widely separated reference stations. The estimated error (correction) components are then transmitted to the user, who applies them to obtain his or her precise position. Several types of government-operated and commercial state-space-domain

WADGPS are currently available on the market. Government-operated state-space-domain WADGPS include four satellite-based augmentation systems (SBAS), namely: (1) wide area augmentation system (WAAS), which was developed by the U.S. Federal Aviation Administration (FAA); (2) European geostationary navigation overlay system (EGNOS), which relies on both GPS and the Russian GLONASS (Chapter 11) systems; (3) India's GPS and GEO augmented navigation (GAGAN) system; and (4) Japan's multifunction transportation satellite (MTSAT)-based satellite augmentation system (MSAS). Similar to WAAS, MSAS relies on the GPS constellation only. The objective of the four augmentation systems is to enhance the GPS SPS to meet the safety requirements in civil aviation. In addition, the four augmentation systems are designed to be compatible and interoperable. Presently, WAAS consists of 25 WAAS reference stations (WRSs), two WAAS master stations, and two geostationary satellites (INMARSAT I-3 satellites) [17]. Data collected at the WRS are used to estimate the differential corrections and the error bounds on them, which are then broadcasted to users as correction messages in a special format called RTCA (Radio Technical Commission for Aeronautics—see Chapter 8) through geostationary satellites [18]. Two different sets of corrections are transmitted at the GPS L1 frequency, namely location-independent parameters (ephemeris and clock errors) and area-specific parameters (ionospheric errors in grid format); both are valid within the WAAS service area [19]. WAAS provides a horizontal positioning accuracy of the order of 1–2m and is available to users with WAAS-enabled receivers at no cost. WAAS achieved IOC status on July 10, 2003. Future WAAS expansion will include more reference stations in Canada and Mexico [17].

Other government-operated state-space-domain WADGPSs, which are not intended for commercial aviation, include the Canada-wide differential GPS (CDGPS) and NASA's global DGPS (GDGPS). The former is a real-time service based on NRCan's GPS corrections (GPSC). It uses data from the CACS GPS tracking network. Once the corrections are generated, they are transmitted to an uplink center in Ottawa via a dedicated communications link, where they are assembled and uploaded to two mobile communication satellites, MSAT-1 and MSAT-2, for broadcast across Canada [20]. Users access the real-time CDGPS service via a compact custom-built L-band receiver, which has an onboard low-cost GPS chip. The receiver can output three different types of user-selectable data sets, namely GPSC corrections, localized RTCM SC-104 corrections, or positions in the National Marine Electronics Association (NMEA) format. The GPSC corrections are transmitted in a modified RTCA format, which is suitable for external dual-frequency GPS receivers. The expected accuracy of the CDGPS service is at the meter level for single-frequency receivers and submeter level for dual-frequency receivers [21]. The CDGPS service achieved IOC status on October 14, 2003 [21]. NASA's GDGPS is a WADGPS system capable of providing seamless global real-time decimeter-level positioning accuracy for dual-frequency receivers, which was developed by JPL [22]. The system uses the carrier-phase and pseudorange measurements of a global network of about 60 reference stations, known as NASA Global GPS Network (GGN), to produce precise GPS satellite orbits and clocks parameters [23]. JPL's Real-Time GIPSY software is used for that purpose. The precise orbit and clock products are then formatted as corrections to the GPS broadcast ephemeris, which are made available to authorized users through various communication means including the Internet [23]. The ionospheric delay is almost eliminated by combining the L1 and L2 measurements, while the tropospheric delay is accounted for by the user (see Chapter 3). Two commercial services with global coverage that are based on JPL's products are currently available, namely C-Nav of C&C Technologies and OmniSTAR XP [24, 25]. Both commercial systems use geostationary satellites to broadcast the corrections over large areas.

7.4 Multisite RTK System

As mentioned in Chapter 5, RTK positioning with a single reference station is limited to a distance of about 15 to 20 km. Beyond this distance limit, the errors at the reference and the rover receivers become less correlated (i.e., dissimilar) and would not cancel out sufficiently through the double differencing [8]. This leads to unsuccessful fixing for the ambiguity parameters, which in turn deteriorates the positioning accuracy. To overcome this limitation, research groups have developed multisite real-time, carrier phase–based RTK positioning [8].

The idea behind multisite RTK positioning is based on using a network of reference stations to create raw GPS measurements for a virtual reference station, which is located very close to the mobile, or the rover, receiver. Once created, the virtual reference station measurements are transmitted to the mobile receiver, where the normal single reference station RTK positioning can be performed. The way, the virtual reference station measurements that are created can be summarized as follows. First, the differential errors between the reference stations within the network are determined, based on their known precise coordinates. The differential errors at any point within

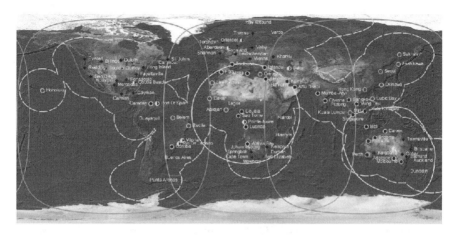

Figure 7.4 Fugro's high performance service reference network. (*Source:* http://www.fugrochance.com/services/navigation.asp.)

the network (e.g., a mobile receiver's location) can then be determined by interpolation. Once the mobile user provides his or her approximate position to the control station, the differential errors at that location are determined. The raw measurements are then created, based on the differential errors and the approximate position of the mobile user [8].

Other forms of multisite RTK positioning have also been developed. The principle, however, remains similar to the virtual reference station technique. An example of a commercial multisite RTK positioning is Fugro's Starfix-High Performance service, which is available for marine applications worldwide (the land version is known as Omnistar-HP) [26]. Figure 7.4 shows the reference stations used by Starfix-HP (OmniSTAR-HP) service. The Starfix-HP (OmniSTAR-HP) service offers positioning solution accuracies of about 10 and 15 cm for the horizontal and vertical components, respectively [26].

References

[1]   Beutler, G., et al., "International GPS Service (IGS): An Interdisciplinary Service in Support of Earth Sciences," 32nd COSPAR Scientific Assembly, Nagoya, Japan, July 12–19, 1998.

[2]   Cheves, M., "No-Cost GPS Observations Available Nationally," *Professional Surveyor*, Vol. 16, No. 2, 1996, pp. 6–10.

[3]   Cheves, M., "CORS: A Case Study of a One-Man GPS Crew," *Professional Surveyor*, Vol. 16, No. 2, 1996, pp. 10–13.

[4]   Duval, R., P. Héroux, and N. Beck, "Canadian Active Control System Delivering the Canadian Spatial Reference System," GIS '96, Vancouver, Canada, March 1996, http://www.geod.nrcan.gc.ca/publications/papers/index_e.php.

[5]   EUREF Permanent Network, http://www.epncb.oma.be/_trackingnetwork/maps.html.

[6]   Radio Technical Commission for Maritime Services, "RTCM Recommended Standards for Differential GNSS Service," Version 3.0, Arlington, VA, February 10, 2004.

[7]   Mueller, T., "Wide Area Differential GPS," *GPS World*, Vol. 5, No. 6, June 1994, pp. 36–44.

[8]   Raquet, J., and G. Lachapelle, "Efficient Precision Positioning: RTK Positioning with Multiple Reference Stations," *GPS World*, Vol. 12, No. 4, April 2001, pp. 48–53.

[9]   Hada, H., et al., "The Internet, Cars, and DGPS: Bringing Mobile Sensors and Global Correction Services On Line," *GPS World*, Vol. 11, No. 5, May 2000, pp. 38–43.

[10] Gurtner, W., "Access to IGS Data," http://igscb.jpl.nasa.gov/components/usage.html.

[11] IGS Products, http://igscb.jpl.nasa.gov/components/prods.html.

[12] The 2001 Federal Radionavigation Plan (FRP), http://www.navcen.uscg.gov/pubs/frp2001/default.htm.

[13] Bisnath, S., et al., "Evaluation of Commercial Carrier Phase-Based WADGPS Services for Marine Applications," *ION GPS/GNSS 2003*, Portland, OR, September 9–12, 2003, CD-ROM.

[14] Abousalem, M., et al., "DGPS Positioning Using WAAS and EGNOS Corrections," *ION GPS 2000*, Salt Lake City, UT, September 19–22, 2000, CD-ROM.

[15] OmniSTAR Frequently Asked Questions, http://www.omnistar.com/faq.html.

[16] OmniSTAR Worldwide DGPS Service, http://www.omnistar.com/about.html.

[17] Walter, T., et al., "Modernizing WAAS," *ION GNSS 17th International Technical Meeting of the Satellite Division*, Long Beach, CA, Sept. 21–24, 2004, CD-ROM.

[18] Misra, P. and P. Enge, *Global Positioning System: Signals, Measurements and Performance*, Lincoln, MA: Ganga-Jamura Press, 2001.

[19] Wide Area Augmentation System (WAAS), http://gps.faa.gov/FAQ/index.htm.

[20] Garrard, G., et al., "Canada-wide DGPS Service—A National GPS Correction Service for Canadians," GIS2002 (Geotec event) Toronto, Ontario, Canada. April 8–11, 2002, http://www.cdgps.com/e/techpapers.htm.

[21] CDGPS service, http://www.cdgps.com.

[22] Muellerschoen, R. J., et al., "Decimeter Accuracy—NASA's Global DGPS for High-Precision Users," *GPS World*, Vol. 12, No. 1, January 2001, pp. 14–20.

[23] The NASA Global Differential GPS, http://www.gdgps.net.

[24] C-Nav Globally Corrected GPS, http://www.cctechnol.com.

[25] OmniSTAR, Inc., http://www.omnistar.com/news/announce4.html.

[26] Lapucha, D., et al., "Comparison of the Two Alternate Methods of Wide Area Carrier Phase Positioning," *ION GNSS 17th International Technical Meeting of the Satellite Division*, Long Beach, CA, Sept. 21–24, 2004, CD-ROM.

8

GPS Standard Formats

Since individual GPS manufacturers have their own proprietary formats for storing GPS measurements, it can be difficult to combine data from different receivers. A similar problem is encountered when interfacing various devices, including the GPS receiver. To overcome these limitations, a number of research groups and agencies have developed standard formats for various user needs. Brief descriptions of some GPS (GNSS)-related standard formats and the agencies that developed them are given in Table 8.1. This chapter discusses the most widely used standard formats, namely, RINEX, SP3, RTCM SC-104, and NMEA 0183.

Table 8.1
Common GPS (GNSS)-Related Standard Formats

Format	Description	Agency
RINEX	GPS and GLONASS observations, meteorological data, and navigation files	IGS (igscb.jpl.nasa.gov)
BINEX	Binary exchange format for GPS, GLONASS, and SBAS data—mainly for research purposes	University Navstar Consortium (UNAVCO) (www.unavco.org)
SP3-c	GPS and GLONASS orbit solutions	IGS
SINEX	Station position and velocity solutions	IGS
IONEX	Ionospheric TEC grid products	IGS

Table 8.1 (continued)

Format	Description	Agency
Tropo SINEX	Tropospheric zenith path delay products	IGS
ANTEX	Antenna calibrations	IGS
RTCM SC-104	Differential and kinematic GNSS corrections and data	RTCM (www.rtcm.org)
RTCA SC-159	Differential corrections and the error bounds on them	RTCA (www.rtca.org)
NMEA 0183	GNSS positioning solution, time, and other information related to GNSS and other sensors	NMEA (www.nmea.org)

8.1 RINEX Format

To save storage space, proprietary (nonstandard) formats developed by GPS receiver manufacturers are mostly binary, which means that they are not directly readable when displayed [1]. This creates a problem when combining data (in the postprocessing mode) from different GPS receivers. To overcome this problem, a group of researchers has developed an internationally accepted data exchange format [1]. This format, known as the RINEX format, is in the standard ASCII format (i.e., readable text). Although a file in the ASCII format is known to take more storage space than a file in the binary format, it provides more distribution flexibility.

A RINEX file is a translation of the receiver's own compressed binary files. A draft version of the RINEX format was introduced in 1989, followed by a number of updates to accommodate more data types (e.g., GLONASS data) and other purposes [1]. The current RINEX version 2.11 defines seven different RINEX files, each containing a header and a data section [2]: (1) GNSS observation data file, (2) GPS navigation message file, (3) meteorological data file, (4) GLONASS navigation message file, (5) geostationary satellites (GPS signal payloads) navigation message file, (6) satellite and receiver clock data file, and (7) SBAS broadcast data file. The record, or line length, of all RINEX files is restricted to a maximum of 80 characters. One of the main characteristics of RINEX version 2.11 is the inclusion of the new GPS and Galileo (the European-owned global navigation satellite system currently under development, see Chapter 11) measurements.

At the time of writing (February 2006), a new RINEX version 3.0 has just been published, which contains significant changes to the RINEX observation file (existing navigation and meteorological files are unchanged). Version 3.0 defines five different RINEX files, each containing a header and a data section [3]: (1) GNSS observation data file, (2) GPS navigation message file, (3) meteorological data file, (4) GLONASS navigation message file, and (5) SBAS broadcast data file. One of the main characteristics of RINEX version 3.0 is the inclusion of a more flexible and detailed definition of the observation coding to accommodate the new GPS and Galileo observables. In addition, the 80-character restriction in the line length of the observation records has been removed in Version 3.0.

For the majority of GPS users, the first three files are the most important and therefore are the only ones discussed here. Interested readers can find further details in [2, 3]. The recommended naming convention for RINEX files is "ssssdddf.yyt." The first four characters, "ssss," represent the station name; the following three characters, "ddd," represent the day of the year of the first record; the eighth character, "f," represents the file sequence number (or character) within the day. For a daily observation file that is generated, the character "f" is assigned a value of zero. On the other hand, for an hourly observation file, "f" is a character between "a" and "x" (i.e., "f = a" for the first hour 00h–01h, "f = b" for the second hour 01h–02h, …, "f = x" for the 24th hour 23h–00h). The file extension characters "yy" and "t" represent the last two digits of the current year and the file type, respectively. The file type takes the following symbols: "O" for the GNSS observation file, "N" for the GPS navigation file, "M" for the meteorological data file, "G" for the GLONASS navigation file, "L" for the future Galileo navigation file, and "H" for the SBAS payload navigation message file. For example, a file with the name "abcd033e.06o" is an hourly observation file for a station "abcd," which was observed on February 2, 2006, at 04h.

The observation file contains in its header information that describes the file's contents, such as the station name, antenna information, the approximate station coordinates, number and types of observations, observation interval in seconds, the time of first observation record, and other information. The types of observations from GPS include carrier-phase measurements (L) in units of cycles (L1, L2, and L5), pseudorange measurements (C) in units of meters (C/A-code and L2C), pseudorange measurements (P) in units of meters (P1 and P2), and Doppler frequencies (D) on L1 and L2 in units of Hertz (D1 and D2). The raw signal strength is represented in the header by the signal-to-noise ratio (SNR). The GPS time frame is used for the GPS files, while the UTC (Soviet Union) time frame is used for GLONASS files. A mixed GPS/GLONASS file must include a time system identifier in its header. The header section may contain some optional records such as the leap seconds. The last 20 characters of each record (i.e., columns 61 to 80) contain textual descriptions of that record. The last record in the header section must be "END OF HEADER." Figure 8.1 shows an example of a RINEX observation file for dual-frequency data.

The data section is divided into epochs; each contains the time tag of the observation—the received-signal receiver time, in the GPS time frame (i.e., not UTC) for GPS files, the number and list of satellites, the various types of measurements in the same sequence as given in the header, and the signal strength. Other information, such as the loss of lock indicator, is also

```
     2.10          OBSERVATION DATA    G (GPS)              RINEX VERSION / TYPE
teqc  2005Sep1                        20060105 01:03:35UTCPGM / RUN BY / DATE
Linux 2.4.20-8|Pentium IV|gcc|Linux|485/DX+             COMMENT
ALGO CACS-ACP    883150    ALGONQUIN PARK  ONT   CANADA  MARKER NAME
40104M002                                              MARKER NUMBER
-Unknown-              GEOD. SURVEY, NATURAL RESOURCES CANADA OBSERVER / AGENCY
1103                  AOA BENCHMARK ACT    3.3.32.2N   REC # / TYPE / VERS
385                   AOAD/M_T           NONE          ANT # / TYPE
    918129.4000 -4345071.2000  4551977.8000            APPROX POSITION XYZ
         0.1000       0.0000        0.0000             ANTENNA: DELTA H/E/N
     1     1                                           WAVELENGTH FACT L1/2
     7    L1    L2    C1    P2    P1    S1    S2        # / TYPES OF OBSERV
    30.0000                                            INTERVAL
L1 PHASE CENTRE   .110m ABOVE ARP                     COMMENT
L2 PHASE CENTRE   .128m ABOVE ARP                     COMMENT
where ARP is the Antenna Reference Point for HI measurement COMMENT
    P1 =     P1  TurboRogue; =    Y1  Benchmark        COMMENT
    L1 = L1(CA)                                        COMMENT
    P2 =     P2  TurboRogue; =    Y2  Benchmark        COMMENT
    L2 = L2(P2) TurboRogue; = L2(Y2) Benchmark         COMMENT
SNR is mapped to RINEX snr flag value [0-9]            COMMENT
L1 & L2: min(max(int(snr_dBHz/6), 0), 9)              COMMENT
  2005     1     5     0     0    0.0000000      GPS   TIME OF FIRST OBS
                                                      END OF HEADER
 05  1  5  0  0  0.0000000  0  7G29G 5G27G10G 2G25G21
 -15755340.507 5 -13055877.715 5  21125054.825   21125053.082   21125053.974
       40.500          39.200
 -18825154.978 7 -14558941.808 5  21225000.991   21224999.522   21225000.235
       44.500          41.200
   -377390.013 2   -294059.499 2  25147350.855   25147349.176   25147349.524
       17.000          12.500
 -21547000.974 8 -15789856.820 7  20405704.413   20405702.813   20405703.314
       48.400          45.500
 -11115502.897 5  -8552198.507 5  22593290.197   22593285.859   22593288.884
       34.900          32.300
 -14309141.922 5 -11149971.793 5  21753938.988   21753938.735   21753938.897
       38.000          34.400
  -5958804.889 5  -5422441.295 4  23578515.903   23578514.141   23578515.900
       30.800          25.900
 05  1  5  0  0 30.0000000  0  7G29G 5G27G10G 2G25G21
 -15820351.751 5 -13105755.584 5  21113883.714   21113881.895   21113882.895
       40.500          38.700
```

Figure 8.1 Example of a RINEX observation file for dual-frequency data.

included in the data section. The data section may optionally contain the receiver clock offset (i.e., from true GPS time) in seconds (see Figure 8.1). Satellites of the different systems are differentiated via a system identifier "s" followed by a two-digit satellite number "nn" (i.e., "snn," where "s" is set to "G" or blank for GPS, "R" for GLONASS, "S" for geostationary signal payload, or "E" for Galileo). The two-digit satellite number "nn" represents the PRN (GPS and Galileo), slot number (GLONASS), or PRN-100 (GEO).

A more compact form of RINEX observation file format was developed recently by Y. Hatanaka, which cuts the information redundancy to reduce the file size [4]. The Hatanaka format is commonly known as the compact RINEX (CRINEX) format. The IGS is currently using the CRINEX to store its RINEX observations files. Hatanaka-compressed observation files are named in the same manner shown earlier for the RINEX observations files, except that the character "O" is replaced by a "D." Software to convert observation files from RINEX to CRINEX (and vice versa) was developed Y. Hatanaka, which is available at ftp://terras.gsi.go.jp/software.

The navigation message file contains the satellite information as described in Chapter 3. In its header, the navigation message contains information such as the date of the creation of the file, the name of the agency conducting the campaign, and other relevant information. Similar to the observation file, the last record in the header section of the navigation file must be "END OF HEADER." Optionally, the header section may contain additional information, such as the parameters of the ionospheric model for single-frequency users (Chapter 4). In addition, almanac parameters relating GPS time and UTC and the leap seconds may optionally be included in the header section of the navigation message. The first record in the data section contains the satellite PRN number, the time tag, and the satellite clock parameters (bias, drift, and drift rate). The subsequent records contain information about the broadcast orbit of the satellite, the satellite health, the GPS week, and other relevant information (see Figure 8.2).

The meteorological file contains time-tagged information such as the temperature (in degrees Celsius), the barometric pressure (in millibars), and the humidity (in percent), which are measured at the observation (tracking) site. The meteorological file starts with a header section containing the observation types (e.g., pressure), the sensors-related information, the approximate position of the meteorological sensor, and other related information. As with the other files, the last record in the header section must be "END OF HEADER." The data section contains the time tags (in GPS time) followed by the meteorological data arranged in the same sequence as specified in the header (see Figure 8.3).

```
     2               NAVIGATION DATA                    RINEX VERSION /
TYPE
CCRINEXN V1.6.0 UX  CDDIS                 06-JAN-06 18:24  PGM / RUN BY /
DATE
IGS BROADCAST EPHEMERIS FILE                              COMMENT
    0.7451E-08 -0.1490E-07 -0.5960E-07  0.1192E-06        ION ALPHA
    0.9011E -0.6554E -0.1311E  0.4588E            ION BETA
    0.000000000000E 0.177635683940E-14    61440     1357 DELTA-UTC: A0,A1,T,W
    14                                                    LEAP SECONDS
                                                          END OF HEADER
  4 06  1  6  0  0  0.0 0.109004788101E-03 0.121644916362E-10 0.000000000000E
     0.142000000000E 0.206562500000E 0.512557064345E-08 0.314143593410E
     0.112131237984E-05 0.741062243469E-02 0.525638461113E-05 0.515376007843E
     0.432000000000E0.122934579849E-06 0.295502331551E0.614672899246E-07
     0.951741079082E 0.275468750000E 0.117914838392E0.833963309348E-08
    -0.857178562071E-10 0.100000000000E 0.135600000000E 0.000000000000E
     0.280000000000E 0.000000000000E0.605359673500E-08 0.398000000000E
     0.424818000000E 0.400000000000E 0.000000000000E 0.000000000000E
 14 06  1  6  0  0  0.0-0.214227475226E-04 0.568434188608E-12 0.000000000000E
     0.640000000000E 0.200937500000E 0.398802326003E-08 0.215926881124E
     0.114180147648E-05 0.238276377786E-02 0.857934355736E-05 0.515360133743E
     0.432000000000E 0.279396772385E-07-0.124358776019E 0.111758708954E-07
     0.984633999193E 0.227656250000E0.199752235783E0.792461580634E-08
     0.157506560781E-09 0.000000000000E 0.135600000000E 0.000000000000E
     0.200000000000E 0.000000000000E0.931322574616E-08 0.640000000000E
     0.431940000000E 0.400000000000E 0.000000000000E 0.000000000000E
                                          ′
                                          ′
                                          ′
```

Figure 8.2 Example of a RINEX navigation file.

```
     2             METEOROLOGICAL DATA                   RINEX VERSION /
TYPE
ACSMET              NRCan                 06-JAN-06 00:16  PGM / RUN BY /
DATE
                                                          COMMENT
ALGO CACS-ACP    883160   ALGONQUIN PARK  ONT  CANADA     MARKER NAME
40104M002                                                 MARKER NUMBER
     3    PR    TD    HR                                   # / TYPES OF
OBSERV
VAISALA             PTB-100A                     0.1     PR SENSOR
MOD/TYPE/ACC
YSI                 44212                         0.2     TD SENSOR
MOD/TYPE/ACC
VAISALA             HMP-35A                       3.0     HR SENSOR
MOD/TYPE/ACC
     0.0            0.0          0.0            200.9     PR SENSOR POS XYZ/H
                                                          END OF HEADER
  06 01 06 00 00 29  983.0   -6.0   88.6
  06 01 06 00 05 22  983.0   -6.0   88.4
  06 01 06 00 10 21  983.1   -6.1   87.8
  06 01 06 00 15 33  983.1   -6.1   86.5
  06 01 06 00 20 19  983.1   -6.1   86.1
  06 01 06 00 25 19  983.2   -6.1   86.0
  06 01 06 00 30 34  983.2   -6.1   85.7
  06 01 06 00 35 19  983.2   -6.1   85.7
  06 01 06 00 40 19  983.2   -6.1   86.2
  06 01 06 00 45 28  983.2   -6.1   86.9
  06 01 06 00 50 19  983.3   -6.2   87.1
  06 01 06 00 55 22  983.3   -6.1   87.2
                                          ′
                                          ′
                                          ′
```

Figure 8.3 Example of a RINEX meteorological file.

Most GPS receiver manufacturers have developed postprocessing software packages that accept GPS data in the RINEX format. Most of these packages are also capable of translating the GPS data in the manufacturer's proprietary format to the RINEX format. However, users should be aware that some software packages change the original raw observations in the translation process (e.g., smoothing the raw pseudorange measurements).

8.2 SP3 Format

As discussed in Chapter 4, several institutions produce precise orbital (ephemeris) data to support applications requiring high-accuracy positioning. The IGS, for example, offers three types of precise orbits, namely final, rapid and ultrarapid (see Chapter 7). To facilitate the exchange of such precise orbital data, the U.S. NGS developed the SP3 format, which later became the international standard [5]. SP3 is an acronym for Standard Product #3, which was originally introduced as SP1 in 1985. More recently, an extended version of SP3, version "c," was introduced. It contains accuracy information about the clock corrections and the observed versus predicted parts of the IGS ultrarapid orbit files [6]. Future versions will use other letters in alphabetical order (i.e., "d," "e," and so forth). The SP3-c file is an ASCII file that contains information about the precise orbital data (in the ITRF reference frame) and the associated satellite clock corrections. The line length of the SP3-c files is restricted to 80 columns (characters). All times are referred to the GPS time system in the SP3-c data standards.

A precise ephemeris file in the SP3-c format consists of two sections: a header and data. The header section consists of 22 lines (see Figure 8.4). The first line starts with the version symbols (#c) and contains information such as the Gregorian date and time of day of the first epoch of the orbit, and the number of epochs in the ephemeris file. Line 2 starts with the symbols (##) and shows the GPS week number, the seconds of the week, the epoch interval, and the modified Julian day (JD). JD is the number of days that have elapsed since 12 noon (UT) on January 1, 4713 B.C., in the proleptic Julian calendar. Modified JD = JD − 2400000.5 (i.e., starts at midnight). Lines 3–7 start with the symbol (+) and show the total number of satellites (on line 3) as well as list the satellites by their respective identifiers (similar to the RINEX "snn" shown earlier). For example, G01 means GPS PRN01. Lines 8–12 start with the symbols (++) and show the accuracy exponents for the satellites shown on lines 3–7. The meaning of the accuracy exponent (ae) is explained as follows: the standard deviation of the orbital error (in the entire file) for a particular satellite = 2^{ae} mm. For example, as shown in Figure 8.4 (lines 3 and 8), satellite G02 (PRN 02) has an accuracy exponent of 4, which means that the standard deviation of its orbital error is 2^4 = 16 mm or 1.6 cm. Lines 13–18 of the SP3-c header are reserved for additional parameters (current and future modification), while lines 19–22 are used freely for comments. Current additional parameters on line 13 include a file type descriptor, which is represented by left-justified, two-character field (e.g., "G " for GPS-only files, "M " for mixed files, "R " for GLONASS-only files, "L " for LEO-only files, and "E " for Galileo-only files), and a time system indicator (either "GPS" for GPS time or "UTC"). Also, line 15 contains two floating-point base numbers (e.g., 1.2500000 and 1.02500000), which are used to compute the standard deviations of individual orbital and clock components with a better resolution (see an example in Figure 8.4).

The data section of the precise ephemeris in the SP3-c format starts at line 23, which contains the data and the time of the first record (epoch). In fact, this is the same time shown in the first line of the header section. Subsequent lines contain the satellite coordinates and the satellite clock data (and their standard deviations) for the current epoch. Each line is assigned for a particular satellite and starts with the character "P," which means a position line. The character "P" is followed by the satellite identifier (represented in

```
#cP2006   1  6  0  0  0.00000000       96 ORBIT IGb00 HLM    IGS
## 1356 432000.00000000   900.00000000 53741 0.0000000000000
+    29    G01G02G03G04G05G06G07G08G09G10G11G13G14G15G16G17G18
+         G19G20G21G22G23G24G25G26G27G28G29G30  0  0  0  0  0
+          0  0  0  0  0  0  0  0  0  0  0  0  0  0  0  0  0
+          0  0  0  0  0  0  0  0  0  0  0  0  0  0  0  0  0
+          0  0  0  0  0  0  0  0  0  0  0  0  0  0  0  0  0
++         3  4  3  3  3  3  3  3  3  3  3  3  3  3  4  3  3
++         4  3  3  3  3  3  4  3  3  4  3  3  0  0  0  0  0
++         0  0  0  0  0  0  0  0  0  0  0  0  0  0  0  0  0
++         0  0  0  0  0  0  0  0  0  0  0  0  0  0  0  0  0
++         0  0  0  0  0  0  0  0  0  0  0  0  0  0  0  0  0
%c G  cc GPS ccc cccc cccc cccc cccc ccccc ccccc ccccc ccccc
%c cc cc ccc ccc cccc cccc cccc cccc ccccc ccccc ccccc ccccc
%f  1.2500000  1.025000000  0.00000000000  0.000000000000000
%f  0.0000000  0.000000000  0.00000000000  0.000000000000000
%i    0    0    0    0     0     0     0     0          0
%i    0    0    0    0     0     0     0     0          0
/* FINAL ORBIT COMBINATION FROM WEIGHTED AVERAGE OF:
/* cod emr esa gfz jpl mit ngs sio
/* REFERENCED TO IGS TIME (IGST) AND TO WEIGHTED MEAN POLE:
/* CLK ANT Z-OFFSET (M): II/IIA 1.023; IIR 0.000
```

Figure 8.4 Example of header section of an SP3-c file.

the same way as in the header section), the x, y, and z coordinates of the satellite in kilometers, and the satellite clock correction in microseconds. Columns 62–69 contain 3 two-digit exponents representing the standard deviations of the satellite coordinates in units of millimeters, while columns 71–73 contain a three-digit exponent representing the standard deviation of the clock correction in units of picoseconds (see Figure 8.5). To compute the actual standard deviation of each component, the floating-point base number for the orbital components on line 15 of the header section are used. For example, if the floating-point base number is 1.25 (Figure 8.4) and the two-digit exponent for the y-coordinate of a particular satellite at a particular epoch is 12 (Figure 8.5), then the standard deviation of the y-coordinate is 1.25^{12} = 15 mm. The standard deviation of the clock correction is computed in a similar manner. Subsequent epochs will have the same structure, and the file ends with the symbol "EOF."

It should be pointed out that satellite coordinates and clock correlation values may optionally be included in a line that begins with the characters

```
/* CLK ANT Z-OFFSET (M): II/IIA 1.023; IIR 0.000
*  2006  1  6  0  0  0.00000000
PG01 -14695.886725  16079.133302 -14966.422729    29.174461 11 11  8 144
PG02  21785.750595 -12536.618452   8042.967318   -22.840386 12 12 12 172
PG03 -11544.524213  20867.307780  11218.033678    64.303629 14  7  9 185
PG04  26083.058776  -5396.997669  -2560.441710   109.000000 12 12  9 130
PG05  -9358.053161 -20719.587676 -14059.605228   457.875362 13  9 10 160
PG06  -9163.622105 -18300.877344  17072.981429   189.890396 10 10  6 147
PG07  15986.390081  -5118.237907 -20132.111843   463.989264 12 10  7 153
PG08  24817.635650    552.717657   9832.794221   -52.503177 10 13 12 159
PG09   2258.542463 -16474.312714 -21181.439348     2.416274 15 10 11 183
PG10   8041.562437 -12754.720534  21763.955811    76.621428 12 11 11 151
PG11   -408.006131  17590.449465 -19900.119350   296.878737 14 11  9 177
PG13   8289.816388  16887.897414  18699.271712    33.301522 12 11 10 178
PG14 -15239.319308   1933.616553 -21590.439465   -21.420122 11 13  9 160
PG15 -24416.798941  -2185.438121  10503.946506   545.444814 10 11  8 170
PG16 -14227.781365   7030.227144  21387.838879    20.639602 12 13 12 167
PG17  15477.560215    238.779965 -21601.157106    19.811330 10  9  7 151
PG18 -19381.958295 -18312.542518  -1621.573285  -213.070048  9 10 14 178
PG19  -6467.522707  25684.047442  -1663.589474   -24.205539 14 11  9 175
PG20   9501.537983  21198.854484 -12978.262330   -36.057555 12 10  9 148
PG21 -18461.310514  -8297.247988  17488.499073   165.307385  9 11  8 129
PG22 -21198.591203  -8572.074886 -13349.118832    60.141659 11 11 13 180
PG23   1076.224055  24604.984441   9652.618758   155.820702 13  9 10 177
PG24  21588.422398   5136.949609 -14525.433848    75.123935  8  8 11 154
PG25 -24399.153428  10609.963683   1028.052545  -128.400235 10 13 14 176
PG26   8171.908531 -24761.474690   2533.917516   -10.090699  8  7 11 144
PG27  18441.830578   7123.124805  18497.702485    30.292098 12 12 12 169
PG28  21842.252686  11066.596004 -10452.815417    35.676380 14 14 10 159
PG29  10454.561208 -23088.105488   7768.957938   487.701729  9  8 12 142
PG30 -16776.422901 -20005.957206  -5467.214385    11.515172 10 10 12 155
*  2006  1  6  0 15  0.00000000
PG01 -14391.908280  14242.090015 -16966.682156    29.176185 11 11  8 145
                                           ,
                                           ,
                                           ,
```

Figure 8.5 Example of data section of an SP3-c file.

"EP," following the satellite coordinates and clock record. In some cases, satellite velocity values and the rate of clock corrections are mixed with this information. To handle this, the position and clock correction record will be on one line, followed by a line containing the velocity and the rate of clock correction record for the same satellite. The line containing the velocity record starts with the letter "V." In addition, correlation values for velocity and the rate of clock correction may optionally be included in a line that begins with the characters "EV," following the velocity and the rate of clock correction record. More details about these optional records can be found in [6].

8.3 RTCM SC-104 Standards for DGPS Services

The RTCM is an advisory organization that was established in 1947 to investigate issues related to maritime telecommunications. Special Committee No. 104 (SC-104) was established in 1983 to develop and recommend standards for transmitting conventional differential (i.e., pseudorange) corrections to GPS users. A draft Version 1.0 of the recommendations was published in November 1985, which contained 16 tentative messages, pending field tests and operational experience [7]. On January 1, 1990, RTCM SC-104 Version 2.0 was released to replace the draft recommendations of the earlier Version 1.0. Among other updates, Version 2.0 included an increase in the number of messages from 16 to 64 and a revised preamble bit pattern to distinguish the new version from the earlier draft [7]. To support RTK applications, new messages were added to a new Version 2.1 of the RTCM SC-104 standards, which was issued on January 3, 1994. Some tentative messages were also fixed in that version. Following this, a new version of the RTCM SC-104, Version 2.2, was introduced on January 15, 1998, which added some tentative messages to support the differential GLONASS and combined GPS/GLONASS operations [8]. Other messages were also added for datum transformations, timing, and higher resolution coordinates of the reference station antenna phase center. More recently, Version 2.3 of the standards was issued on August 20, 2001, which was followed by the latest Version 3.0 on February 10, 2004. Both versions 2.3 and 3.0 are maintained by the RTCM as the current standards.

In Version 2.3 of the RTCM SC-104 standards, a number of messages were added to improve the potential accuracy of RTK operations and to provide guidance on the use of Loran-C as a medium for broadcasting differential GNSS corrections [9]. This version also included an enhancement to the reference material and a new radiobeacon almanac message to support multiple reference stations. Table 8.2 shows a list of the various message types in Version 2.3 of the RTCM SC-104 standards. Message types 14, 15, 17, 20, and 21, which were previously tentative, were fixed in Version 2.3. In addition, message types 23, 24, and 27 were added as new tentative messages in the same version.

Table 8.2
RTCM SC-104 Version 2.3 Message Types

Message Type No.	Current Status	Title	Message Type No.	Current Status	Title
1	Fixed	DGPS corrections	20	Fixed	RTK carrier phase corrections
2	Fixed	Delta DGPS corrections	21	Fixed	RTK/high-accuracy pseudorange corrections
3	Fixed	GPS reference station parameters	22	Tentative	Extended reference station parameters
4	Tentative	Reference station datum	23	Tentative	Antenna type definition record
5	Fixed	GPS constellation health	24	Tentative	Antenna reference point (ARP)
6	Fixed	GPS null frame	25–26	–	Undefined
7	Fixed	DGPS radiobeacon almanac	27	Tentative	Extended beacon almanac
8	Tentative	Pseudolite almanac	28–30	–	Undefined
9	Fixed	GPS partial correction set	31	Tentative	Differential GLONASS corrections
10	Reserved	P-code differential correction	32	Tentative	Differential GLONASS reference station parameters
11	Reserved	C/A-code L1, L2 delta corrections	33	Tentative	GLONASS constellation health

Table 8.2 (continued)

Message Type No.	Current Status	Title	Message Type No.	Current Status	Title
12	Reserved	Pseudolite station parameters	34	Tentative	GLONASS partial differential correction set
13	Tentative	Ground transmitter parameters	35	Tentative	GLONASS beacon almanac
14	Fixed	GPS time of week	36	Tentative	GLONASS special message
15	Fixed	Ionospheric delay message	37	Tentative	GNSS system time offset
16	Fixed	GPS special message	38–58	–	Undefined
17	Fixed	GPS ephemeris	59	Fixed	Proprietary message
18	Fixed	RTK uncorrected carrier phases	60–63	Reserved	Multipurpose usage
19	Fixed	RTK uncorrected pseudoranges	64	–	Not reported

The RTCM SC-104 messages are not directly readable; they are streams of binary digits, zeros and ones. Each RTCM SC-104 Version 2.x (x = 0, 1, 2, or 3) message or frame consists of "N + 2" words, each is 30 bits long; where N represents the number of words containing the actual data within the message and the remaining two words represent a two-word header at the beginning of each message. The size of N varies, depending on the message type and the message content (e.g., the varying number of satellites in view of the reference station). The word size and the parity check algorithm are the same as those of the GPS navigation message.

RTCM SC-104 Version 2.x messages contain information such as the pseudorange correction (PRC) for each satellite in view of the reference receiver, the rate of change of the pseudorange corrections (RRC), and RTK- and differential GLONASS-related data and corrections (see Table 8.2). Of interest to conventional (i.e., meter-level) real-time DGPS users are message types 1 and 9. Both contain the PRC and the RRC corrections. However, message type 1 contains the corrections for all the satellites in view of the reference station, while message type 9 packs the corrections in groups of three.

This leads to a lower latency for message type 9 compared with message type 1, which was useful in the presence of SA. The disadvantage of using message type 9, however, is that the reference station requires a more stable clock. In this section, we will focus our discussion on the structure of message type 1, which is commonly used in conventional real-time DGPS operations.

Figure 8.6 shows the structure of a message type 1, where five satellites were visible at the reference station. The first word of the header starts with an 8-bit preamble, which is a fixed sequence 01100110. Following the preamble are a 6-bit message type identifier and a 10-bit reference station ID. The last 6 bits of this word and of all other words are assigned for parity, which checks for any error. The second word starts with a 13-bit modified z-count, a time reference for the transmitted message, followed by a 3-bit sequence number for verifying the frame synchronization. The length of frame is assigned bits 17–21 and is used to identify the start of the next frame. Bits 22–24 define the reference station health status (e.g., a code of "111" means that the reference station is not working properly). The actual data set for all the satellites is contained in the remaining words. Each satellite requires a total of 40 bits for the correction message, distributed in the following sequence: (1) scale factor, S (1 bit); (2) user differential range error, UDRE (2 bits); (3) satellite ID (5 bits); (4) pseudorange correction, PRC (16 bits); (5) range-rate correction, RRC (8 bits); and (6) issue of data (8 bits). The scale factor is used to scale the PRC/RRC, while the UDRE is a measure of the uncertainty in the PRC. However, the issue of data identifies the GPS navigation message that the reference station used to calculate the satellite position and clock offset. Evidently, there will be cases where the required number of words is not an exact integer number. For example, as shown in Figure 8.6, if the number of satellites in the message is five, 16 bits are required to fill the frame. Similarly, if the number of satellites is four, 8 bits are required to fill the frame. To avoid confusion with the preamble, the fill uses alternating ones and zeros (i.e., 1010101010101010 or 10101010).

To obtain the DGPS corrections, the transmitted messages must be decoded and converted to 30-bit long words (strings of zeros and ones). Once this is done, parity checks should be performed and then the DGPS corrections information can be extracted according to Figure 8.6. Figure 8.7 shows a "real" example showing four type 1 messages; each represents the DGPS corrections for all the satellites in view of the reference station at a particular epoch. Figure 8.8 shows how the first word of the header information is decoded. Most GPS receivers support Version 2 (some support Version 3 as well) of the RTCM SC-104 standards, which allow the use of different receivers in the real-time mode. It should be noted, however,

1	2	3	4	5	6	7	8	9	10	11	12	13	14	15	16	17	18	19	20	21	22	23	24	25	26	27	28	29	30
Preamble (01100110)								Message type						Reference station ID										Parity					
Modified Z-count													Sequence number			Length of frame					Station health			Parity					
S	UDRE		Satellite ID					Pseudorange correction																Parity					
Range-rate correction								Issue of data						S	UDRE		Satellite ID							Parity					
Pseudorange correction																Range-rate correction								Parity					
Issue of data								S	UDRE		Satellite ID					Pseudorange correction (upper bite)								Parity					
Pseudorange correction (lower bite)								Range-rate correction								Issue of data								Parity					
S	UDRE		Satellite ID					Pseudorange correction																Parity					
Range-rate correction								Issue of data						S	UDRE		Satellite ID							Parity					
Pseudorange correction																Range-rate correction								Parity					
Issue of data								Fill																Parity					

Figure 8.6 Structure of a five-satellite message type 1.

Y~}o}□_X~Cp|_TSVA@VL_OJ]K|C~_^@gJuDAA@UVwIAjl`BAoOxc~WZSc^jTQB`TAPMRUq/
Y~}Oj□_^~GX}}FXtrOL`@T\Js□□A`deZuJo~~□bxHv~Ez`zyEHiFl@GbCuHeW}m]zc^`jtH
fABhH@`oA`n}_k~or}i|`dv\gC|AP_YfDulp|□tri|□Ef`bwEBiFD@Jt\gD\hBJi_|b_Uoå
Y~}wE@`kA^X}m__tqOLZ□Sjmw□_AhWe`QjF`□GjiVorU_`\JGFxGl@y]~Jwv~~□jiHvnjqÐ

Figure 8.7 Example of raw RTCM SC-104 corrections (message type 1).

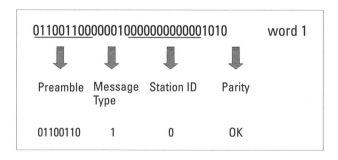

Figure 8.8 First-word decoding example.

that not all the differential-ready GPS receivers could output the RTCM standards.

RTCM SC-104 standards Version 2.x was found to contain some weaknesses [10]. Among them is the parity scheme described earlier, which was shown to be wasteful of bandwidth and not independent from word to word. In addition, the actual integrity of the message was not as high as it should be, despite the fact that the parity was assigned 6 bits. Furthermore, it was found that the 30-bit words are somewhat cumbersome to deal with. To overcome these weaknesses, an initial release of the new RTCM SC-104 standards, Version 3.0, was developed [10]. The new version of the standards consists mainly of message types that support GPS and GLONASS RTK operations (Table 8.3), which involve transmitting too much information (e.g., code and carrier-phase observables and other parameters). The messages in Version 3.0 are permanent and presented in a database format. Future changes, if any, will require the development of new messages. The parity scheme in Version 3.0 uses a 24-bit cyclic redundancy check (CRC) at the end of a variable-length message, which operates on all bits in the message [11]. To ensure efficiency and flexible messaging structure, Version 3.0 is designed using a layered approach as shown in Table 8.4.

Table 8.3
RTK Message Groups

Group Name	Subgroup Name	Message Type
Observations	GPS L1	1001
		1002
	GPS L1/L2	1003
		1004
	GLONASS L1	1005
		1006
	GLONASS L1/L2	1007
		1008
Station coordinates		1009
		1010
Antenna description		1011
		1012
Auxiliary operation information		1013

Source: RTCM SC-104 standards, Version 3.0.

Table 8.4
Various Layers of RTCM SC-104 Standards, Version 3.0

No.	Layer name	Description
1	Application layer	Defines how the messages can be applied for different end-user applications
2	Presentation layer	Describes the messages, data elements, and data definitions
3	Transport layer	Defines the frame architecture for sending or receiving the messages
4	Data link layer	Defines how the message data stream is encoded on the physical layer
5	Physical layer	Defines how the message data is conveyed at the electrical and mechanical level

While the RTCM SC-104 maintained backward compatibility among versions 2.x, the latest Version 3.0 is not compatible with earlier versions of the standards [10, 11]. However, some legacy message types, including differential GPS and GLONASS corrections, radiobeacon almanac, ASCII messages, GPS ephemeris information, and proprietary messages, will be incorporated into Version 3.0 [11].

8.4 NMEA 0183 Format

NMEA was founded in 1957 by a group of electronics dealers to strengthen their relationships with electronic manufacturers [12]. In 1983, with input from the manufacturers and private and governmental organizations, the association adopted the NMEA 0183 as a format for interfacing marine electronic devices. It has been updated several times; the latest release, as of this writing, Version 3.01, appeared in January 2002 [13]. The NMEA 0183 standards are data streams in the ASCII format, transmitted at a rate of 4,800 bps, from a talker to a listener (one-way), where a talker is a device that sends data to other devices (e.g., a GPS receiver) and a listener is a device that receives data from another device (e.g., a laptop computer interfaced with the GPS receiver) [13]. A high-speed addendum to Version 3.01 of NMEA 0183 is also available, which operates at a rate of 38.4 K-baud.

The NMEA 0183 data streams may include information on position, datum, water depth, and other variables. The data is sent in the form of sentences, each starting with a dollar sign "$" and terminating with a carriage return-line feed "<CR><LF>"; the dollar sign "$" is followed by a five-character address field, which identifies the talker (the first two characters) and the format and data type (the last three characters). The data fields are separated by "," delimiter. A null field (i.e., a field with zero length) is used when the value of the field is unavailable or unreliable. The last field in any sentence follows a checksum delimiter character "*" and is a checksum field. The maximum total number of characters in any sentence is 82 (i.e., a maximum of 79 characters between the starting delimiter "$" and the terminating "<CR><LF>"). A number of these sentences are dedicated to GNSS, including GPS and GLONASS systems, while the remaining sentences support other devices such as Loran-C, echo sounders, gyroscopes, and others [13]. Table 8.5 shows common GNSS-related NMEA sentences, which are defined in the current version of NMEA 0183.

Our discussion will be restricted to one sentence only, which is commonly used in practice—the GGA, or GPS fix data. This sentence represents the time, position, and solution-related information. Figure 8.9 shows the general structure of the GGA sentence, while Table 8.6 explains the terms of the sentence.

Table 8.5
Common GNSS-Related NMEA Sentences

NMEA Sentence	Description
ALM	GPS almanac data
GBS	GNSS satellite fault detection
GGA	GPS fix data
GMP	GNSS map projection fix data
GNS	GNSS fix data
GRS	GNSS range residuals
GSA	GNSS DOP and active satellites
GST	GNSS pseudorange error statistics
GSV	GNSS satellites in view

```
$GPGGA,hhmmss.ss,llll.ll,a,yyyyy.yy,a,x,xx,x.x,x.x,M,x.x,M,x.x,xxxx*hh<CR><LF>
$GPGGA,115417.00,4338.123456,N,07938.123456,W,1,10,01.1,095,M,,M,999,0000
```

Figure 8.9 General structure of a GGA sentence.

Table 8.6
Explanation of GCA Sentence Terms

Symbol	Explanation
$	Start of sentence delimiter
GP	Talker identifier (GPS in this case)
GCA	Data identifier (GPS fix data in this case)
,	Data field delimiter
hhmmss.ss	Time of position in UTC system (hoursminutesseconds.decimal). Decimal digits are variable.
llll.ll	Latitude (degreesminutes.decimal). Decimal digits are variable.
a	N/S (North or South)
yyyyy.yy	Longitude (degreesminutes.decimal). Decimal digits are variable.
a	E/W (East or West)
x	GPS quality indicator (0 = Fix not available or invalid)
	(1 = Point positioning with C/A-code)
	(2 = DGPS with C/A-code)
	(3 = Point Positioning with P-code)
	(4 = RTK with ambiguity parameters fixed to integer values)
	(5 = RTK with float ambiguity parameters)
	(6 = Estimated [dead reckoning] mode)
	(7 = Manula input mode)
	(8 = Simulator mode)
xx	Number of satellites used in producing the solution
x.x	Horizontal Dilution of Precision (HDOP)
x.x	Altitude above geoid (Orthometric height)
M	Meters (units of Orthometric height)
x.x	Geoidal Height above the WGS84 ellipsoid (geoid-ellipsoid separation)
M	Meters (units of Geoidal Height)
x.x	Age of Differential GPS data in seconds (time since last RTCM message type 1 or 9 was received; null filed when Differential GPS mode is not used).
xxxx	Differential reference station ID (range 0000-1023)
*	Cheksum delimiter character
hh	Checksum field (last field in the sentence)
<CR><LF>	Sentence terminator

Most GPS receivers available on the market support the NMEA 0183 standards. However, not all receivers with the NMEA 0183 port output all the GPS-specific messages. In addition, some GPS receiver manufacturers may slightly change the standard format. However, they typically provide software to interpret the data sentence.

References

[1] Gurtner, W., "RINEX: The Receiver-Independent Exchange Format," *GPS World*, Vol. 5, No. 7, July 1994, pp. 49–52.

[2] Gurtner, W., and L. Estey, "RINEX: The Receiver Independent Exchange Format Version 2.11," http://igscb.jpl.nasa.gov/components/formats.html.

[3] Gurtner, W., and L. Estey, "RINEX: The Receiver Independent Exchange Format Version 3.0," February 1, 2006, http://igscb.jpl.nasa.gov/components/formats.html.

[4] Hatanaka, Y., "Compact RINEX Format, Version 1.0," January 3, 1998, http://igscb.jpl.nasa.gov/components/formats.html.

[5] Remondi, B., "Extending the National Geodetic Survey Standard Orbit Formats," NOAA Technical Report NOS 133 NGS 46, 1989.

[6] Hilla, S., "The Extended Standard Product 3 Orbit Format (SP3-c)," September 5, 2002, http://igscb.jpl.nasa.gov/components/formats.html.

[7] Radio Technical Commission for Maritime Services, www.rtcm.org.

[8] Radio Technical Commission for Maritime Services, "RTCM Recommended Standards for Differential GNSS Service," Version 2.2, Alexandria, VA, January 15, 1998.

[9] Radio Technical Commission for Maritime Services, "RTCM Recommended Standards for Differential GNSS Service," Version 2.3, Alexandria, VA, August 20, 2001.

[10] Radio Technical Commission for Maritime Services, "RTCM Recommended Standards for Differential GNSS Service," Version 3.0, Arlington, VA, February 10, 2004.

[11] Kalafus, R. M., and K. Van Dierendonck, "The New RTCM SC-104 Standard for Differential and RTK GNSS Broadcasts," ION GPS/GNSS 2003, Portland, OR, September 9–12, 2003.

[12] National Marine Electronic Association, "NMEA 0183 Standards for Interfacing Marine Electronics," Version 3.0, New Bern, NC, July 2000.

[13] National Marine Electronic Association, "NMEA 0183 Standards for Interfacing Marine Electronics," Version 3.01, New Bern, NC, January 2002.

9

GPS Integration

GPS has found its way into many applications, mainly as a result of its accuracy, global availability, and cost-effectiveness. Unfortunately, there exist some situations in which part of the GPS signal may be obstructed to the extent that the GPS receiver may not "see" enough satellites for positioning. Examples of these situations are positioning in urban canyons and deep open-pit mining. This signal-obstruction problem is successfully overcome by integrating GPS with other complementary systems. Integration of GPS and other complementary systems is also beneficial in the cases of significant attenuation. This chapter introduces the most common integrated systems, namely GPS/Loran, GPS/laser range finder, GPS/dead reckoning, GPS/inertial navigation system, GPS/pseudolite, and GPS/cellular system.

9.1 GPS/Loran-C Integration

Loran is a terrestrial radio-navigation system, which provides positioning and timing services for users within its coverage area [1, 2]. Loran is an acronym for long-range navigation. The first generation, known as Loran-A, was developed in early 1940s. Its successor, Loran-C, provides longer range and greater accuracy [3]. A Loran-C system consists of ground transmitting stations separated by hundreds of kilometers, which are organized into chains. Typically, a Loran chain encompasses a master station and two or more (up to five) secondary stations [4]. A minimum of two secondary stations is required to obtain a position fix in the normal mode of operation (see Figure 9.1). However, a user with a precise clock can obtain a position fix with two stations only. Such a positioning mode is known as the Rho-Rho mode. Secondary stations are designated by the letters U, W, X, Y, and Z. For example, two secondary stations would be designated by (Y and Z), while three secondary stations would be designated by (X, Y, and Z), and so on.

Loran-C operates in the band 90–110 kHz. The master and secondary stations within a chain are synchronized and transmit a series of radio pulses centered on 100 kHz at precise time intervals. The pulses are transmitted in sequence; the master station transmits its pulses (a total of nine) first and then each of the secondary stations transmits its pulses (a total of eight) in turn, according to the designation letter. To illustrate this mechanism, consider the case of a Loran chain consisting of a master and two secondary stations (Figure 9.2). First, the master station transmits a series of nine pulses; the first eight are spaced at 1 ms, while the ninth pulse is transmitted 2 ms after the eighth. After some time, known as the emission delay for Y (ED_Y), the first secondary Y transmits its group of eight pulses at an interval of 1 ms. Following this, the secondary station Z transmits its group of eight pulses (again at an interval of 1 ms) after the master by a time ED_Z. To ensure proper sequence of pulse transmission, ED_Z is selected to be longer than

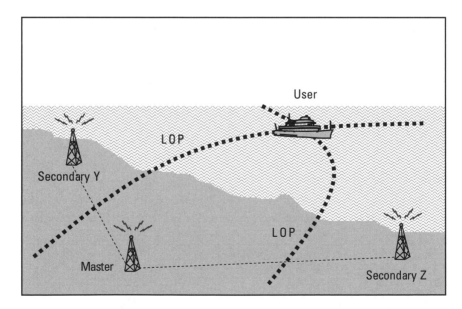

Figure 9.1 Principle of the Loran system.

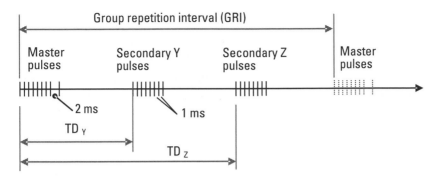

Figure 9.2 Loran-C pulse pattern.

ED_Y. Once the last (i.e., the second in our example) secondary station has transmitted its pulse group, the master station transmits again and a new cycle begins. The time it takes the chain to complete one cycle is known as group repetition interval (GRI), which varies from 50 to 100 ms [4].

A Loran-C receiver operating within a chain's coverage area measures the time difference (TD) between the arrival of the pulse groups from the master station and each of the secondary stations. Loran receivers are programmed to identify and track the arrival time of the third cycle zero crossing within the pulses (Figure 9.3). Since the locations of the transmitting stations and the emission delays are known, the measured TD can be converted to distance, which defines a hyperbolic line of position (LOP). The intersection of two (or more) such LOPs defines the receiver's horizontal position (Figure 9.1). Because of the poor vertical dilution of precision (VDOP) (see Chapter 4) of the Loran system, an accurate estimate of the receiver's height component is not possible.

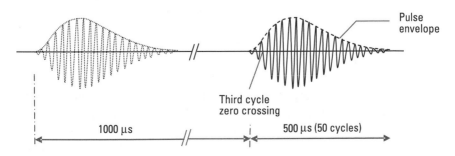

Figure 9.3 Individual Loran-C pulse (not to scale).

According to [3], the predictable (i.e., absolute) accuracy of the Loran-C system is on the order of 0.25 nm (2drms) or better. This accuracy limitation is caused by the presence of both random errors (mainly noise) and biases (mainly signal propagation effects). Loran signal propagation is affected by the atmosphere through which the signal travels and the conductivity of the surface over which the signal passes (i.e., sea or land). Land surfaces have lower conductivity than seawater, which means that the propagation speed of the Loran signal is lower over land than it is over seawater. Atmospheric error is accounted for through the introduction of a correction term known as the primary phase factor (PF). Another correction term, known as the secondary phase factor (SF), is used to account for the propagation error due to traveling over seawater. A third correction term, known as the additional secondary phase factor (ASF), is introduced to account for the delay due to the land conductivity. Of the three corrections, the ASF is the most complex [1].

In examining the characteristics of the Loran system, we find that it has strengths and weaknesses. Fortunately, the strengths and weaknesses are very different from those of GPS. While the Loran signal is affected by atmospheric noise (e.g., thunderstorms and lightning bursts), GPS is not [5]. On the other hand, Loran is a relatively high-power system, while GPS is a low-power system. Additionally, Loran broadcasts at a low frequency (100 kHz), which means that it occupies a very different portion of the frequency spectrum than GPS. As such, jamming the Loran signal would be more difficult. An important characteristic of the Loran system is that its signal is available in urban canyons and under foliage, while the GPS signal may be blocked. These characteristics suggest that the two systems complement each other. In other words, the two systems can be integrated to overcome their weaknesses and take advantage of their strengths. One way of integrating GPS and Loran can be done in the measurements level through the integration of GPS pseudoranges and Loran time differences [4]. Figure 9.4 shows a GPS/Loran receiver, which integrates the two systems at the measurement domain. GPS can also be used for cross-chain synchronization of traditional Loran-C. Moreover, GPS can be used to calibrate the Loran propagation errors, both time- and location-dependent.

Because of its potential as a backup to GPS, a number of countries are modernizing the Loran system. A modernized or enhanced Loran (eLoran) transmitters use a time of emission (TOE) control where every transmitter is synchronized with respect to UTC [6]. A Loran reference station can be installed to collect measurements and calculate differential eLoran corrections, which could be transmitted to users in its vicinity to improve the

Figure 9.4 GPS/Loran receiver (*From:* www.reelektronika.nl).

autonomous Loran accuracy. This method is known as differential Loran, which is similar in principle to differential GPS. Reported differential eLoran precision is in the order of 25m at 95 percent probability level [5], which is significantly better than the original Loran autonomous precision.

9.2 GPS/LRF Integration

In areas with heavy tree canopy, GPS receivers will normally lose lock to the GPS satellites. In addition, real-time differential GPS corrections may not be received as well. To overcome these problems, integrated GPS/handheld laser units, or laser range finders (LRFs), were developed [7]. The way the integrated system operates is to set up the GPS antenna in a nearby open area, which allows the GPS system to operate normally without losing lock to the GPS satellites. With the help of a digital compass, a reflectorless handheld laser that is co-located with the GPS receiver can be used to determine the distance and azimuth to the inaccessible points (see Figure 9.5). This operation is commonly known as the offset function. Software residing in the handheld computer helps in collecting both the offset data and the GPS data. At a later time, all the available information is processed using PC software to determine the coordinates of the inaccessible points. Collecting and processing the data may also be done in real time, while in the field, provided that the real-time DGPS corrections can be received (otherwise, GPS will be used in autonomous mode). Once the processing is done, the user can export the output to the required Geographical Information System (GIS) or

Figure 9.5 GPS/LRF integration.

CAD software. This eliminates the need to place the GPS antenna directly on the features to be mapped [8].

GPS/laser integration is an attractive tool, especially for the forestry industry. Tree offsets, heights, and diameters can be measured easily with the laser unit. From a single location, a stationary user in a relatively open area can offset any number of points or features. In this case, the user location will be determined precisely by averaging all the GPS data collected while taking the offset measurements. Other applications of the GPS/laser integration include mapping points under bridges, mapping points on a busy roadway, mapping highway signs, and mapping shore lines, to name a few. GPS/laser integration can be used to map point features, line features, or area features.

9.3 GPS/Dead Reckoning Integration

Another system that has been used to supplement GPS under poor signal reception is the dead reckoning (DR) system. Dead reckoning is a low-cost system, commonly comprising an odometer sensor and a vibration gyroscope. The integrated GPS/DR system is widely used in automatic vehicle location (AVL) applications [9].

DR navigation requires that the vehicle travel-distance (or, equivalently, speed) and direction (heading) be available on a continuous basis. The travel-distance information is obtained from the odometer sensor, while the direction information is obtained from the gyroscope. If the vehicle starts the trip from a known location, the distance and direction information can be used to determine the vehicle location at any time. In other words, assuming that the vehicle is traveling in a horizontal plane, the travel and direction information can be integrated over time to compute the vehicle location (position).

Odometer sensors are already installed in all vehicles, mainly to evaluate their age and whether a service is required. An odometer sensor counts the number of revolutions of the vehicle's wheels, which can be converted to a travel distance through an initial calibration. This conversion is known as the odometer scale-factor determination. One way of determining the scale factor is by driving the vehicle over a known distance. Unfortunately, however, the odometer scale factor changes over time due mainly to wheel slipping and skidding, tire pressure variation, tire wear, and vehicle speed. If left uncompensated, the scale-factor error will accumulate rapidly, causing significant positional error [10].

Vibration gyroscopes (or *gyros*) are low-cost sensors that measure the angular rate (heading rate) based on the so-called Coriolis acceleration (i.e., when an object moves on the surface of a rotating body, it undergoes a force known as Coriolis force). A vibration gyro outputs a voltage that is proportional to the angular velocity of the vehicle. The vehicle's heading rate is obtained by multiplying the output voltage by a scale factor. Similar to the odometer sensors, gyroscopes suffer from error accumulation due to gyro bias and scale-factor instability. A gyro bias is a temperature-sensitive variable error that affects the gyro measurements at all times. As such, a gyro will read a nonzero value even if the angular velocity is zero. It is observable when the vehicle is stationary or when it is moving in a straight line. However, the gyro scale-factor error affects the gyro measurements only when the vehicle is taking a turn. This error could be greatly reduced by taking equal clockwise and counterclockwise rotations [9].

It can be seen that each of the GPS and DR systems suffers from limitations. While the GPS signal may not be available in obstructed areas, the DR system drifts over time causing large positional error. This suggests that an optimal positioning solution may be developed, based on the two positioning systems. Kalman filtering technique is commonly used to integrate the two systems [10]. With the integrated system, GPS helps in controlling the drift of the DR components through frequent calibration, while DR becomes the main positioning system during the GPS outages. As such, the performance of the integrated system will be better than either system alone.

9.4   GPS/INS Integration

There exists a number of applications that require high-accuracy positioning in obstructed areas and/or under high-dynamic conditions. Examples of these applications are deep open-pit mining and airborne mapping (see Chapter 10 for details about these applications). As discussed earlier, a major problem with GPS is its limitation when used in obstructed areas. In addition, a GPS receiver has limited dynamic capabilities. As mentioned in Section 2.8, GPS signal obstruction and high receiver dynamics can cause temporary signal losses, or cycle slips. Furthermore, when orientation (or attitude) is required, such as for airborne or ocean-floor mapping (see Chapter 10), GPS provides insufficient accuracy. To overcome these limitations, GPS can be integrated with a relatively environment-independent system, the inertial navigation system (INS).

Inertial navigation can be defined as the process of finding the position, velocity, and orientation of a vehicle based on the measurements of inertial sensors [11]. Inertial technology has evolved from the complex gimbaled system where sensors are mounted in a set of gimbles to strapdown systems where sensors are mounted directly on the vehicle. A gimbaled system uses mechanical gyroscopes with spinning wheels, while a strapdown system uses more advanced inertial sensor technology such as optical gyroscopes [12].

An inertial sensor, also known as the inertial measurement unit (IMU), is a device consisting of a cluster of accelerometers and gyroscopes, supporting electronics components (to operate the sensors), and power supply. The IMU often contains an internal microprocessor to compensate for some of the sensors' drifts [11]. When mounted on a moving object, the accelerometers measure the object's acceleration plus the gravitational force, while the gyroscopes measure the object's rotational motion (rate of rotation), both with respect to an inertial frame. As discussed in the previous section, inertial sensors produce electrical signals, which through comparison with precise values (i.e., calibration) can be interpreted to indicate specific values for rotation and/or accelerations. Usually, an inertial system contains three mutually perpendicular accelerometers; each measures the acceleration in one direction. The gyroscopes are used to track the directions in which the accelerometers are pointing. By successive integration of the inertial sensor's measurements with respect to time, the changes in the object's velocity and position can be determined. If an initial position and velocity information are available, the system becomes an autonomous navigation system providing three-dimensional position and velocity information [13]. An important process that must be carried out prior to the start of inertial navigation, however, is the determination of the initial orientation of the IMU measurement axes with respect to the axes of the reference frame. Such a process is known as initial alignment [14].

In addition to being a relatively environment-independent system, an inertial system provides accuracy as high as that of GPS for the short period of time following the initialization [13]. Moreover, inertial systems provide very high update rates compared with GPS. However, a major drawback of the inertial system is that it suffers from drift if left unaided for a long period of time. In particular, the performance of the gyroscopes limits the overall performance of the inertial system. To overcome such limitations, INS can be integrated with GPS. In fact, integrating GPS and INS overcomes the limitations of both systems, as the two systems complement each other [13]. While GPS provides the initialization and the correction to the inertial system, the latter bridges the GPS gaps when the satellite signal is blocked or temporarily lost. GPS/INS integration is commonly done in either of two modes, namely, loose coupling or tight coupling mechanisms. Loosely coupled integration is carried out in the solution (position) domain, while tightly coupled integration is carried out in the raw measurements domain. Tightly coupled integration requires extensive computations as compared with loosely coupled integration. However, it results in a nearly optimal integration solution. A third level of GPS/INS integration, which is presently under development, is known as the ultratight (or deep) integration. Deep integration combines GPS signal tracking and INS, which has a number of advantages such as fast signal reacquisition following a brief interruption and improved multipath resistance [11]. Similar to the GPS/DR, the Kalman filtering technique is commonly used for GPS/INS integration [10].

It should be pointed out that, although the new generation of inertial sensors have many advantages over old technology, their size, cost, and weight have limited the use of inertial systems for certain applications. In recent years, MEMS-based technology has evolved, with a number of unique characteristics. MEMS systems are miniature devices, which comprise one or more micromachined components or subsystems [15]. In addition to their small size, MEMS systems are inexpensive, light, and consume very low power compared with conventional sensors. As a result, a wide range of applications is expected to benefit from MEMS technology, including, for example, the automotive industry, medical instrumentation, telecommunications, and environmental monitoring. MEMS systems are commonly fabricated using silicon, which possesses significant electrical and mechanical advantages over other materials. A critical stage in MEMS fabrication is the packaging process, which constitutes about 75 percent to 95 percent of the total cost of a MEMS device [15]. So far, as a result of the diversity of MEMS applications, no standards exist for MEMS packaging. Large packaging will destroy the small-size advantage that MEMS technology possesses [15].

The focus of MEMS-based inertial sensors' research is on accelerometers and gyroscopes [16]. According to [15], micromachined accelerometers were first demonstrated in 1979 at Stanford University. This was followed by a steady increase in the unit-volume market, which was accompanied by a decrease in the unit cost. There exist different designs for silicon accelerometers, including the simple pendulum [12]. Micromachined gyroscopes, on the other hand, operate based on the principle of the Coriolis effect. Although not yet matured, MEMS accelerometers and gyroscopes have exceeded the expectations in their performance. One of the major challenges that have yet to be overcome is the high noise of the output of MEMS inertial sensors compared with that of conventional technology.

9.5 GPS/Pseudolite Integration

One of the fastest growing applications of GPS is open-pit mining. The use of GPS in open-pit mining can remarkably reduce the cost of various mining operations. The availability of real-time GPS positioning at centimeter-level accuracy has attracted the attention of the mining industry. This is mainly because accurate real-time positioning is a key component that leads to automating the heavy and expensive mining machines. As such, smart mining systems can be developed that not only increase mining safety but also reduce costly labor [19].

Unfortunately, similar to the earlier cases, the satellite signal will be partially blocked as the pit deepens (see Figure 9.6). As such, in deep open-pit mining, GPS alone cannot be used reliably for mining positioning. One promising system that can augment GPS to ensure high-accuracy positioning at all times is the pseudolite (short for pseudo-satellite) system. A pseudolite is a ground-based electronic device that transmits a GPS-like signal (code, carrier frequency, and data message), which can be acquired by a GPS receiver. Unlike GPS, which uses atomic clocks onboard the satellites, pseudolites typically use low-cost crystal clocks to generate the signal [18].

The addition of pseudolite signals improves both system availability and geometry. The number and locations of the pseudolites can be optimized to ensure the best performance of the system. The vertical dilution of precision, in particular, can be improved dramatically, which leads to improved accuracy for the height component. Another advantage of using

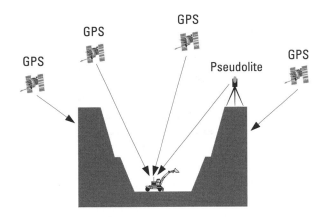

Figure 9.6 GPS/pseudolite integration.

the pseudolites is that, being ground-based transmitters, their signals are not affected by the ionosphere. However, pseudolites suffer from a number of drawbacks that must be overcome to ensure high-accuracy positioning. The first is known as the near-far problem, which results from the variation in the received pseudolite signal power as the receiver-pseudolite distance changes. The closer the receiver is to the pseudolite transmitter, the higher the signal power will be, and vice versa. This problem does not exist with GPS-only positioning, as the received GPS signal power remains almost constant, because the satellite-receiver distance does not change significantly. Consequently, in GPS/pseudolite integration, if a given pseudolite signal is much stronger than the signals from the other pseudolites and GPS satellites, it may overwhelm these other signals and jam the receiver. This is what is known as the near problem. However, if the pseudolite signal is much weaker, the receiver may not be able to track it, which is known as the far problem. Transmitting the pseudolite signal in short pulses with a low duty cycle may, however, minimize the effect of the near-far problem [18].

The use of inaccurate clocks to generate the pseudolite signal causes a synchronization error in the sampling time. This error will cause a range error, even if double differences are formed. A possible solution to this problem is through the use of a content-free data message of a master pseudolite. Another problem that requires the pseudolite user's attention is the multipath error. Pseudolite multipath error occurs as a result of reflected signals from objects surrounding the antennas of both the receiver and the transmitter. Some researchers have suggested the use of patterned antennas as a feasible way of reducing the multipath effect. Unlike GPS-only positioning, where ephemeris errors do not affect the position solution significantly, errors in the pseudolite coordinates will be propagated into the solution, causing large positioning errors. This is caused by the relatively short receiver-pseudolite separation [19]. Careful calibration of the pseudolite antenna location solves this problem.

It should be pointed out that the application of the integrated GPS/pseudolite system is not limited to deep open-pit mining. Such an integrated system has been successfully used in precise aircraft landing, deformation monitoring, and other applications. Being similar in principle to GPS, pseudolite-only positioning has the potential of being a major contender for the system of the future for indoor applications, such as underground mining (see Figure 9.6). However, a challenge to overcome is the pseudolite location (sitting) problem.

9.6 GPS/Cellular Integration

Cellular communication technology is becoming widely accepted throughout the world. Both the number of subscribers and the cellular coverage areas are increasing continuously. In addition, more advanced digital cellular coverage is on the rise, allowing voice and data to be mixed seamlessly. This makes the cellular system very attractive to a number of markets, including emergency 911 (E-911), AVL, and RTK GPS.

A major limitation with the current standalone cellular system, however, is its ability to precisely determine where a call was originated [20]. Although this limitation is not critical for applications like RTK GPS, it is of utmost importance for other applications, such as E-911 (E-112 in Europe) and AVL. In the United States, for example, about one-third of all emergency 911 calls come from cellular phones. Of these, nearly a quarter cannot describe their location precisely, which makes it very difficult for an operator to effectively send out assistance. As such, the U.S. Federal Communications Commission (FCC) made it mandatory that wireless emergency 911 callers must be located with an accuracy of 125m (67 percent probability level) or better [20].

To meet the FCC location requirement, wireless network operators can either use a network-based location or a handset-based location. Most network-based caller location systems employ either the time-difference of arrival (TDOA) approach or the angle of arrival (AOA) approach to determine the caller's location. The former measures the differences in the arrival times of an emergency 911 signal at the cell sites or base stations. The caller's location can be determined if the signal is received at a minimum of three base stations. Obviously, time synchronization is essential with this technique, which can be ensured by equipping each cell site with a GPS timing receiver. The second technique, the AOA, uses phased-array antennas to compute the angles at which the signal arrives at the base stations. A minimum of two sites is required to compute the caller's location with this method. As both the TDOA and AOA methods have advantages and drawbacks, some network operators combine the two methods [20].

Handset-based location technology integrates GPS with cellular communication through the installation of a GPS chipset in the handset of the wireless phone. With SA being turned off, this technology would locate the wireless emergency 911 callers with an accuracy that exceeds the FCC requirement by a factor of ten. Unlike network-based technology, handset-based location technology is very simple to implement and does not require the installation of additional equipment at the base stations (e.g., GPS timing receivers). However, one of the drawbacks of the handset-based location technology is that the GPS signal received inside buildings is too weak. This limitation can largely be overcome using the so-called assisted GPS (A-GPS), technology. As the name indicates, a GPS receiver (in the handset of the wireless phone) operating within a cellular telephone network receives some information from the network assistance server, which helps the receiver to track and utilize weaker signals. Such information, which may be sent to the user as short message service (SMS) messages, includes GPS satellite ephemeris and clock data, initial receiver position, and time estimate [21]. The availability of such information helps increase the receiver sensitivity, which in turn allows the receiver to track weaker signals (such as those received inside buildings) that otherwise would not be available. In addition, with A-GPS, the time to first fix (TTFF) is substantially reduced, which enhances the wireless location-based services (LBS).

The advances in the wireless communication and caller's location technologies discussed earlier will greatly impact a number of industries. The vehicle navigation market, for example, is expected to greatly benefit from the advances in wireless communication, location, and Internet technologies (see Section 10.11 for details about vehicle navigation). Currently, vehicles use complex systems that integrate location technology with in-car computer navigation systems containing electronic digital road maps and other related information. Clearly, the in-car system will not be aware of any real-world changes in the navigation system's database (e.g., a change in the traffic direction). With the availability of wireless Internet service, however, an up-to-date database residing at a central location could be accessed by drivers, eliminating the need for a complex in-car computer navigation system.

Furthermore, with the availability of a precise location system, drivers could customize the information they need according to their locations, such as turn-by-turn navigation, traffic information, and local weather conditions. This method is simple, cost-effective, and flexible, and it has the potential to become the way of the future.

References

[1]   Bowditch, N., "The American Practical Navigator," Bicentennial Edition, NIMA Pub. No. 9, 2002. CD-ROM.

[2]   Prasad, R. and M. Ruggieri, *Applied Satellite Navigation Using GPS, Galileo, and Augmentation Systems*, Norwood, MA: Artech House, 2005.

[3]   FRP, U.S. Federal Radionavigation Plan, 2001.

[4]   Enge, P., and F. van Graas, "Integration of GPS and Loran-C," in *Global Positioning System: Theory and Applications, Vol. II*, B. Parkinson and J. Spilker (Eds.), American Institute of Aeronautics and Astronautics, Washington, D.C., 1996.

[5]   Macaluso, J., "Loran-C Modernization—Achievements and Plans," *ION NTM 2005*, San Diego, CA, January 24–26, 2005.

[6]   LORADD series, Integrated GPS/eLoran Receiver, version 1.01, reelektronika, 2005, http://www.reelektronika.nl.

[7]   Laser Technology, Inc., "Survey Laser for Forestry," Presentation, http://www.lasertech.com/ downloads.html.

[8]   Ashtech, Inc., "Reliance Field Asset Management Tools," Magellan Corporation, Santa Clara, CA, 2001.

[9]   Madhukar, B. R., et al., "GPS-DR Integration Using Low Cost Sensors," *Proc. ION GPS-99*, 12th Intl. Technical Meeting, Satellite Division, Institute of Navigation, Nashville, TN, September 14–17, 1999, pp. 537–544.

[10] Kaplan, E., *Understanding GPS: Principles and Applications*, Norwood, MA: Artech House, 1996.

[11] Titterton, D., and J. Weston, *Strapdown Inertial Navigation Technology*, 2nd ed., American Institute of Aeronautics and Astronautics, Reston, VA, 2004.

[12] Lawrence, A., *Modern Inertial Technology*, 2nd ed., New York: Springer-Verlag, 1998.

[13] May, M. B., "Inertial Navigation and GPS," *GPS World*, Vol. 4, No. 9, September 1993, pp. 56–66.

[14] Chatfield, A., *Fundamentals of High Accuracy Inertial Navigation*, American Institute of Aeronautics and Astronautics, Reston, VA, 1997.

[15] Maluf, N., *An Introduction to Microelectromechanical Systems Engineering*, Norwood, MA: Artech House, 2000.

[16] Tung, S., "Position Paper: An Overview of MEMS Inertial Sensors," Department of Mechanical Engineering, University Of Arkansas, Fayetteville, AR, 2000.

[17] El-Rabbany, A., "Mining Positioning," *Proc. Smart Systems for Mineral Resources Workshop*, Toronto, Ontario, February 14, 2001.

[18] Cobb, S., and M. O'Connor, "Pseudolites: Enhancing GPS with Ground-Based Transmitters," *GPS World*, Vol. 9, No. 3, March 1998, pp. 55–60.

[19] Wang, J., et al., "Integrating GPS and Pseudolite Signals for Position and Attitude Determination: Theoretical Analysis and Experiment Results," *Proc. ION GPS 2000*, 13th Intl. Technical Meeting, Satellite Division, Institute of Navigation, Salt Lake City, UT, September 19–22, 2000, pp. 2252–2262.

[20] Driscoll, C., "Wireless Caller Location Systems," *GPS World*, Vol. 9, No. 10, October 1998, pp. 44–49.

[21] LaMance, J., and J. DeSalas, "Assisted GPS—A Low-Infrastructure Approach," *GPS World*, Vol. 13, No. 3, March 2002, pp. 46–51.

10

GPS Applications

GPS has been available for civil and military use for more than two decades. This period has witnessed the creation of numerous new GPS applications. Because it provides high-accuracy positioning in a cost-effective manner, GPS has found its way into many industrial applications, replacing conventional methods in most cases. For example, with GPS, machineries can be automatically guided and controlled. This is especially useful in hazardous areas, where human lives are endangered. Even some species of birds are benefiting from GPS technology, as they are monitored with GPS during their migration season. This way, help can be presented as needed. This chapter describes how GPS is being used in land, marine, and airborne applications.

10.1 GPS for the Utility Industry

Accurate and up-to-date maps of utilities are essential for utility companies. The availability of such maps helps electric, gas, and water utility companies to plan, build, and maintain their assets.

GPS, integrated with a geographical information system (GIS), provides a cost-effective, efficient, and accurate tool for creating utility maps. GIS is a computer-based tool capable of acquiring, storing, manipulating, analyzing, and displaying spatially referenced data [1]. Spatially referenced data is data that is identified according to its geographic location (e.g., features such as streets, light poles, and fire hydrants are linked by geography). With the help of GPS, locations of features such as gas lines can be accurately collected, along with their attributes (such as their conditions and whether or not a repair is needed). Once the information is collected, a GIS stores it as a collection of layers in the GIS database (see Figure 10.1), which can then be used to create updated utility maps.

In situations of poor GPS reception, such as in urban canyons, it might be useful to use integrated GPS and LRF systems [2]. This integrated system is an efficient tool for rapid utility mapping. A GPS receiver remains in the open for the best signal reception, while the LRF measures the offset information (range and azimuth) to the utility assets such as light poles (see Figure 10.2). The processing software should be able to combine both the GPS and the LRF information.

Buried utilities such as electric cables or water pipes can also be mapped efficiently using GPS (Figure 10.2). With the help of a pipe/cable locator attached to the second port of the GPS handheld controller, accurate information on the location and the depth of the buried utility can be collected. This is a very cost-effective and efficient tool, as no ground marking is required.

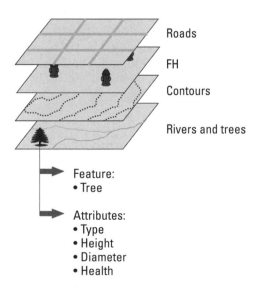

Roads

FH

Contours

Rivers and trees

Feature:
• Tree

Attributes:
• Type
• Height
• Diameter
• Health

Figure 10.1 GIS for utility mapping.

Figure 10.2 GPS for utility mapping.

10.2   GPS for Forestry and Natural Resource

GPS has been applied successfully in many areas of the forest industry. Typical applications include fire prevention and control, harvesting operations, insect infestation, boundary determination, and aerial spraying [3].

With thousands of fires facing the forest services every year, an efficient resource-management system is essential. GPS is a key technology that enables the system operator to identify and monitor the exact location of the resources (Figure 10.3). With the help of GIS and a good communication system, appropriate decisions can be made.

In the past, aerial photography was the only means of providing information on the shape and location of cut blocks before completing harvesting operations. Such information was often lacking in accuracy. With the use of differential GPS, however, this information can be accurately determined in real time.

GPS has also been a very useful tool for wildlife management and insect infestation. Using its precise positioning capability, GPS can determine the locations of activity centers. These locations can be easily accessed using GPS waypoint navigation (see Section 10.15).

Figure 10.3 GPS applications in forestry.

GPS surveying is becoming the preferred method for forest boundary determination. With real-time GPS, a reduction in time and cost for a typical surveying project of up to 75 percent is possible. As discussed in Chapter 9, in case of poor GPS reception under heavy tree canopy, it might be useful to use integrated GPS and LRF systems. Other integrated systems, including GPS/digital barometers and GPS/laser digital videography, have been applied successfully in the forest industry as well.

10.3 GPS for Precision Farming

The ability of DGPS to provide real-time submeter- or even decimeter-level accuracy has revolutionized the agricultural industry [4, 5]. GPS applications in precision farming include soil sample collection, chemical applications control, and harvest yield monitors (Figure 10.4).

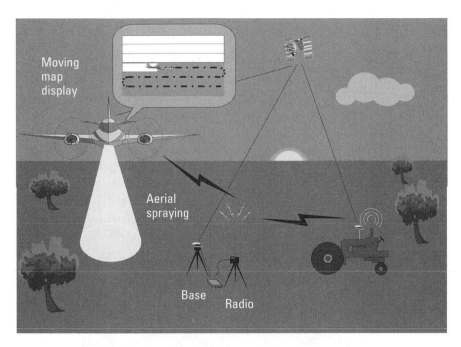

Figure 10.4 GPS for precision farming.

When collecting soil samples, GPS is used to precisely locate the sample points from a predefined grid (Figure 10.4). After testing the soil samples, information such as nitrogen and organic material content can be obtained. This type of information is mapped and used as a reference to guide farmers to efficiently and economically treat soil problems.

When GPS is integrated with an aerial guidance system, the field sprayer can be guided through a moving map display. Based on the sprayer's location, the system will apply the chemicals at the right spots, with minimal overlap, and automatically adjust their rate. This, in addition to increasing productivity, ensures that chemicals and fuel are used efficiently.

GPS is also used to map crop yields. As the DGPS-equipped harvester moves across the field, yield rates are recorded along with DGPS-derived positions. This information is then mapped to show the yield rate.

Easy-to-use integrated systems with only a few buttons are now available on the market. DGPS corrections are available from the government-operated DGPS/beacon service free of charge as well as from a number of commercial services. The user's own base station may be built as well (Figure 10.4).

10.4 GPS for Civil Engineering Applications

Civil engineering works are often done in a complex and unfriendly environment, making it difficult for personnel to operate efficiently. The ability of GPS to provide real-time submeter- and centimeter-level accuracy in a cost-effective manner has significantly changed the civil engineering industry. Construction firms are using GPS in many applications such as road construction, earth moving, and fleet management.

In road construction and earth moving, GPS, combined with wireless communication and computer systems, is installed onboard the earth-moving machine [6]. Designed surface information, in a digital format, is uploaded into the system. With the help of the computer display and the real-time GPS position information, the operator can view whether or not the correct grade has been reached (see Figure 10.5). In situations in which millimeter-level elevation is needed, GPS can be integrated with rotated beam lasers [1].

Figure 10.5 GPS for construction applications.

The same technology (i.e., combined GPS, wireless communications, and computers) is also used for foundation works (e.g., pile positioning) and precise structural placement (e.g., prefabricated bridge sections and coastal structures). In these applications, the operators are guided through the onboard computer displays, eliminating the need for conventional methods [7].

GPS is also used to track the location and usage of equipment at different sites. By sending this information to a central location, GPS enables contractors to deploy their equipment more efficiently. Moreover, vehicle operators can be efficiently guided to their destinations.

10.5   GPS for Monitoring Structural Deformations

Since its early development, GPS has been used successfully in monitoring the stability of structures, an application that requires the highest possible accuracy. Typical examples include monitoring the deformation of dams, bridges, and television towers. Monitoring ground subsidence of oil fields and mining areas are other examples where GPS has been used successfully. In some cases, GPS may be supplemented by other systems such as INS or total stations to work more efficiently. Deformation monitoring is done by taking GPS measurements over the same area at different time intervals [7].

Slow-deforming structures such as dams require submillimeter- to millimeter-level accuracy to monitor their displacement. Although this accuracy level may be achieved with GPS alone under certain conditions, it is not a cost-effective method [7]. To effectively monitor such structures, GPS should be supplemented with geotechnical sensors and special types of total stations.

Bridges, in contrast, are subjected to vibrations caused by dynamic traffic loads. To effectively monitor such cyclic deforming structures, dual GPS receivers should be located at several points with maximum amplitude of cyclic deformation [7]. For example, in monitoring the world's longest suspension bridge (Akashi Bridge in Japan), a GPS receiver is installed at the midpoint of the bridge, while two others are installed at the main towers. Figure 10.6 shows another example in which the Ashtech Z12 dual-frequency receiver is used for monitoring bridge deformation. As the GPS data collection rate is currently limited to 10 Hz, an INS system may supplement the GPS system, in some cases, to monitor the high-frequency portion of structural vibration.

Figure 10.6 GPS for monitoring bridge deformation. (*Courtesy of:* Magellan Corporation.)

10.6 GPS for Open-Pit Mining

Until recently, conventional surveying was the only method available for staking drill patterns and other mining surveying tasks. However, as a result of the harsh mining environment, stakes were often buried or displaced. In addition, drill operators had no precise way of determining the actual drill penetration depth. Likewise, there was no way of monitoring the drill performance in the various geological layers or monitoring the haul trucks in an efficient way. However, more recently the development of modern positioning systems and techniques, particularly RTK GPS, has dramatically improved various mining operations [8, 9]. In open-pit mines, for example, the use of RTK GPS has significantly improved several mining operations such as drilling, shoveling, vehicle tracking, and surveying. RTK GPS provides centimeter-level positioning accuracy and requires only one base receiver to support any number of rovers. As the pit deepens, part of the GPS signal may be blocked by the steep walls of the mine, causing a positioning problem. However, this problem has been successfully overcome by integrating GPS with other positioning systems, mainly the pseudolite system (see Section 9.5) [10].

The mining cycle includes several phases, with ore excavation being one of the most important [11]. Excavating the ore is done by drilling a pre-defined pattern of blast holes, which are then loaded with explosive charges. The pattern of blast holes is designed in such a way that the size of the rock fragmentation is optimized. As such, it is important that the drills are positioned precisely over the blast holes; otherwise, redrilling may be required. An efficient way of guiding the drills is through integrating GPS with a drill navigation and monitoring system consisting of an onboard computer and drilling software. Some systems utilize two GPS receivers, mounted on the top of the drill mast, for precise real-time positioning and orientation of the drill. The designed drill patterns are sent to the onboard computer via radio link and are then used by the integrated system to guide the drill operator to precisely position the drill over blast holes (see Figure 10.7). This is done automatically without staking out. In addition, the onboard computer displays other information, such as the location and depth of each drill hole. This is very important, as it allows the operator to view whether or not the target depth has been reached. In addition, the system accumulates information on the rock hardness and the drill productivity, which can be sent to the engineering office in near real time via a radio link. Such information can be used not only to monitor drill productivity from the engineering office, but also to understand rock properties. This enables better future planning [11].

GPS is also used for centimeter level–accuracy guidance of shoveling operations (Figure 10.7). Shovels are used in loading the ore into the haul trucks, which then transport it and unload it in stockpiles. With an integrated GPS and shovel guidance and monitoring system, elevation control can be automated. With the help of the system display, shovel operators are able to keep the correct grade. This is done automatically without the need for grade control by conventional surveying methods. Similar to the drilling, shoveling productivity can be sent to the engineering office in near real time via a radio link for monitoring and analysis.

In transporting the ore, haul trucks use continuously changing mining roads and ramps. Unless efficiently routed, safety and traffic problems would be expected, which cause an increase in the truck cycle time. The use of GPS, wireless communication, and a computerized dispatch system onboard the haul truck solves this problem efficiently. With the help of a computerized dispatch system, haul trucks can be guided to their destination using the best routes. In addition, the dispatch center can collect information on the status of each haul truck as well as the traffic conditions. Analyzing the traffic conditions is particularly important in devising a more appropriate road design [11].

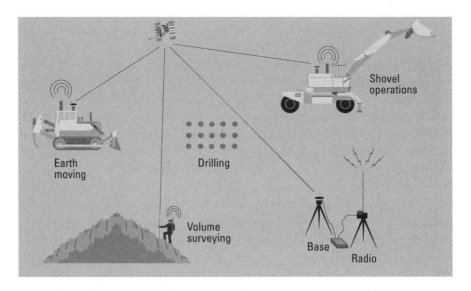

Figure 10.7 GPS use in open-pit mining.

GPS is also used in other phases of the mining cycle (e.g., in checking the coordinates of the individual points and volume surveying). Either the RTK or the non-RTK GPS could be used for these functions (Figure 10.7).

10.7   GPS for Land Seismic Surveying

Oil and gas exploration requires mapping of the subsurface geology through seismic surveying. In land seismic surveys, low-frequency acoustic energy is sent down into the underground rock layers (Figure 10.8). The source of the acoustic energy is often selected to be a mechanical vibrator consisting of a metal plate mounted on a truck. The plate is pressed against the ground and vibrated to produce the acoustic energy. In rough areas, dynamite is still being used as the energy source.

As the acoustic energy (signal) crosses the various underground rock layers, it is affected by the physical properties of the rocks. Portions of the signal are reflected back to the surface by the various layers. The reflected energy can be detected by special seismic devices called geophones, which are laid out at known distances from the energy source along the survey line (Figure 10.8). Upon detecting seismic energy, geophones output electrical signals that are proportional to the intensity of the reflected energy [12]. The

Figure 10.8 GPS for land seismic surveying.

electrical signals are then recorded on magnetic tapes for geophysical analysis and interpretation.

It is clear that unless the positions of the energy source and the geophones are known with sufficient accuracy, the very expensive seismic data becomes useless. GPS is used to provide the positioning information in a standard or a user-defined coordinate system. Integrated GPS/GLONASS and GPS/digital barometer systems have been used successfully in situations of poor GPS signal reception [13]. With the help of GPS, the environmental impacts (e.g., the need to cut trees) as well as the operating cost of seismic surveys have been reduced significantly.

10.8 GPS for Marine Seismic Surveying

Marine seismic surveying is similar in principle to land seismic surveying. That is, a low-frequency acoustic energy is sent down into the subsurface rock layers and is reflected back to the surface to reveal information about the composition of subsurface rocks (Figure 10.9).

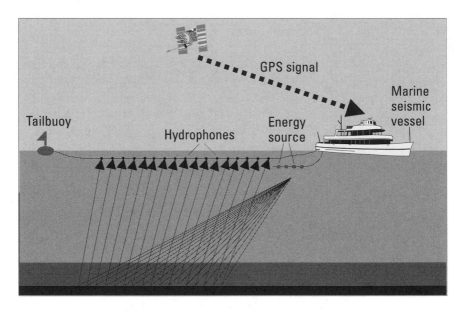

Figure 10.9 GPS for marine seismic surveying.

Different methods are used in marine seismic surveys depending on the water depth. In deep waters, seismic vessels tow seismic cables, known as streamers, which contain devices called hydrophones used for detecting reflected energy. A single vessel will normally tow four to eight parallel streamers, each having a length of several kilometers [12]. The low-frequency acoustic energy is generated using a number of air guns towed behind the vessel at about 6m below the surface. In shallow waters, both the land and the marine methods are used. Ocean bottom cable survey is a relatively new technology that has been used recently for water depths of up to about 200m. In this method, hydrophones and geophones are combined in a single receiver to avoid water column reverberation (Figure 10.9).

To obtain meaningful results, the positions of the energy source and the hydrophones must be known with sufficient accuracy. This can be easily achieved, at low cost, with GPS. Moreover, it is possible to revisit the points precisely with the GPS waypoint feature (see Section 10.15).

As the operation of marine seismic surveys is very expensive, the issue of quality control is essential. To maintain quality control, the seismic industry has suggested the use of two independent positioning systems, with GPS being the primary one [12].

10.9 GPS for Airborne Mapping

GPS alone has been successfully used for topographic mapping of small-size areas. Using either conventional GPS kinematic surveying or GPS RTK, a user takes positions of the points on the ground where the topography changes, which can be used at a later time to produce the topographic map of that area. Even in rough areas, GPS can be mounted on all-terrain vehicles to precisely map those areas. However, there exist situations in which the use of GPS alone becomes time-consuming and/or not cost-effective [14]. Examples include mapping large areas, coastal areas, forests, and inaccessible areas.

Traditionally, mapping large and inaccessible areas was done using classical airborne photogrammetry. With this method, an aircraft-mounted camera is used to capture a sequence of images for the area to be mapped, which after processing construct the map. To be of practical use, the captured images must first be related to the geodetic reference system (e.g., WGS 84), a process known as georeferencing the images. In classical airborne photogrammetry, the georeferencing is done indirectly with the help of a number of ground control stations with known geodetic coordinates and their corresponding image coordinates. In recent years, GPS has been used onboard the aircraft to provide the precise position of the aerial camera as well as the precise time of each aerial exposure (Figure 10.10) [15].

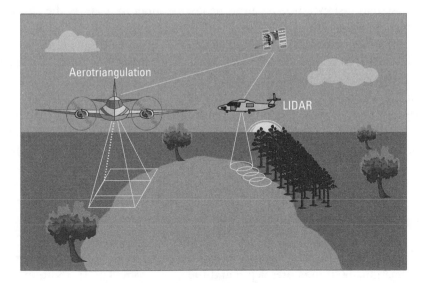

Figure 10.10 GPS for airborne mapping.

The use of GPS in airborne photogrammetry has significantly reduced the required number of ground control points. However, it has not eliminated the necessity for aerial triangulation or the ground control points. This limitation is overcome by augmenting GPS with a high-quality IMU; see Section 9.4 for details about this integration. The integrated GPS/inertial system provides not only the precise position of the imaging sensor but also its orientation (attitude). This allows for the captured images to be directly related to the geodetic reference system without using ground control points. In other words, direct georeferencing of the captured images can be achieved when using an integrated GPS/ inertial system onboard the aircraft. In practice, however, a minimum number of ground control points may be needed for assessing the resulting accuracy [15].

Direct georeferencing using the integrated GPS/inertial system is gaining wide acceptance and is expected to become the standard tool for rapid determination of sensor position and orientation. With recent advances in digital airborne imaging sensors and digital photogrammetric workstations, the integrated GPS/inertial system will enable a fully digital photogrammetric workflow to be developed. With digital imaging sensors, no film development and scanning are required, which further reduces the time and the cost of photogrammetric work. Other applications, such as airborne remote sensing and light detection and ranging (LIDAR), will greatly benefit from the direct georeferencing using the integrated GPS/inertial system. The latter system, LIDAR, uses an airborne laser scanner to measure the altitude of the points above the ground level [16]. Combining the GPS/inertial-based position and orientation of the laser with the measured altitude of the points leads to direct acquisition of accurate digital elevation models. Another advantage of the LIDAR system is that the data can be collected at night as well as under cloudy and high-wind conditions. In addition, the ability of the laser to penetrate to the ground, even in forest areas, makes the airborne laser system attractive to the forest industry. Moreover, the LIDAR system can be used in mapping featureless areas, such as deserts and areas covered by snow and ice [17].

10.10   GPS for Seafloor Mapping

Safe and efficient marine navigation requires, among other factors, accurate information about the water depth and the sea bottom [18]. In addition, the availability of accurate water depth is vital for making use of maximum cargo capabilities. This is especially important for areas with shallow water depth.

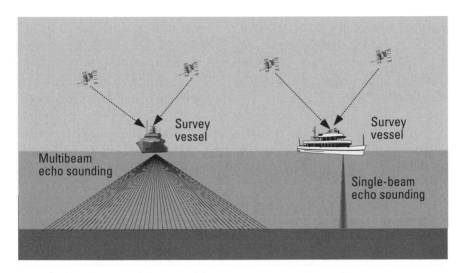

Figure 10.11 GPS for seafloor mapping.

The traditional way of obtaining the water depth was based on the use of a single-beam echo sounder installed on a survey vessel. With this method, the single-beam echo sounder generates a sounding wave (pulse), which is transmitted to the sea bottom and then reflected back to the echo sounder (see Figure 10.11). The water depth is then computed based on the recorded travel time of the sounding pulse and the velocity of the sound in the water [18]. It should be pointed out that the echo sounder uses a hull-mounted device called the transducer to convert the electrical energy into sound energy and vice versa.

To map an area with a single-beam echo sounder, a survey vessel follows preplanned track lines, while the echo sounder generates soundings along the track. Line spacing (the distance between tracks) is selected to provide the best coverage of the area. The accuracy and the reliability of the surveyed depths and locations are verified by supplementing the primary sounding lines with a series of cross lines [18]. This method is characterized by its simplicity. In addition, the echo sounder orientation is not required. A major drawback, however, is that it is time-consuming and does not provide complete coverage of the seafloor.

In recent years, a new technology for seafloor mapping has evolved that combines multibeam echo sounders, GPS, and INS. Multibeam echo sounders employ multiple sounding waves propagating at varying angles, which allow whole swaths of acoustic information to be collected on both sides of the track lines (see Figure 10.11). Unlike single-beam echo sounders, this multibeam technology offers complete coverage of the seafloor with high resolution, provided that the track lines are optimally designed [19]. Optimal line spacing is determined based on the approximate water depth, the footprint of sound, and the bottom profile. GPS waypoint navigation can be employed in the field to ensure that the vessel follows the designed track lines.

Because of their wide swath (usually 150 degrees), multibeam echo sounders require accurate positioning and attitude of the vessel. This is especially important for the outer beams. Integrated GPS/INS is used for this purpose. Some manufacturers have developed an integrated GPS/INS system that utilizes two GPS receivers and antennas. Besides offering accurate positioning and attitude of the vessel, this integrated system estimates the heading of the vessel at high accuracy, regardless of the vessel's dynamics and latitude.

Another state-of-the-art technology that has found wide acceptance within the hydrographic community is the airborne laser bathymetry system. This system operates on the same principle as the land-based airborne laser system (i.e., an aircraft-mounted laser sensor transmits a laser beam, which is partially reflected from the sea surface and from the seafloor). The water depth can then be computed by measuring the time difference between the returns of the two reflected pulses. An accurate three-dimensional seafloor map can be derived from the depth information and the GPS/inertial-based position and orientation of the laser. The major advantages of this method are high productivity and efficiency in mapping difficult areas such as narrow passages. It is, however, limited to shallow water areas (maximum depth about 50m). In addition, it is very sensitive to the water clarity.

10.11   GPS for Vehicle Navigation

When traveling through unfamiliar areas, vehicle drivers often use paper road maps for route guidance. However, besides being inefficient, searching for a destination using a paper map is unsafe, especially in busy areas. A new technology, incorporating GPS with digital road maps and a computer system, has been developed so that route guidance can be obtained electronically with a touch of a button [20]. Figure 10.12 illustrates this concept.

The role of GPS in this technology is to continuously determine the vehicle's location. In obstructed areas, such as urban canyons and tunnels,

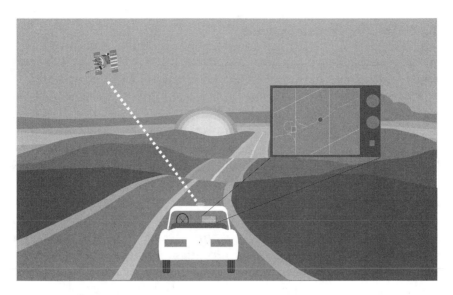

Figure 10.12 GPS for vehicle navigation.

GPS is supplemented by a terrestrial system such as the DR system to overcome the GPS signal blockage. As discussed in Section 9.3, DR is a system that uses the vehicle's odometer and a selection from accelerometers, compasses, and gyros to determine the vehicle's direction and distance traveled. This system is accurate only over a short period of time.

The GPS-determined vehicle location is superimposed on an electronic digital road map, containing in its database digital information such as street names and directions, business listings, airports, attractions, and other related information. Once the driver inputs a destination, the built-in computer finds the best route to reach that destination. Such factors as shortest distance and time to destination, one-way roads, illegal turns, and rush-hour restrictions are all considered in the path finding. Some systems allow the drivers to input other factors such as accident avoidance. The driver usually gets turn-by-turn instructions, with audio and/or visual indications, to the destination. If the driver misses a turn, the system displays a warning message and finds an alternative best route based on the current location of the vehicle. Some manufacturers add cellular systems to provide weather and traffic information and to locate the vehicles in case of emergency. Recent advances in wireless communication technology even make it possible for drivers to remotely access the Internet from their vehicles [21].

10.12   GPS for Transit Systems

Transit system authorities in many countries are faced with a challenging trend of fiscal constraints, which limits their capabilities to expand existing services and to increase ridership. Until recently, transit systems used old technologies such as odometer/compass sensors and signposts for position determination [22]. Odometers are sensors that measure the number of rotations generated by the vehicle's wheels, which are then used to estimate the distance traveled by the vehicle. With the help of a compass, the vehicle's direction of travel can be determined at any time. Combining the measurements from the odometer and the compass, the vehicle's position can be determined with respect to an initial (known) position. Unfortunately, both the odometer and the compass drift over time, which causes significant error in the estimated position. Signposts, in contrast, are radio beacon transmitters that are placed at known locations along the bus routes [23]. Each beacon transmits a low-power microwave signal, which is detected by a receiver on the bus, to account for the odometer's drift error. Unfortunately, this system has a number of limitations, including its incapability of knowing the exact location of a vehicle in between two signposts. In addition, it is not possible to track a vehicle that goes off-route as a result of, for example, a road closure [22].

To overcome the limitations of these systems, transit authorities are integrating a low-cost autonomous GPS system with one or more of these conventional systems. GPS helps in controlling the drift of the conventional systems through frequent calibration. In addition, the vehicle's position can be obtained reliably with GPS if the vehicle goes off-route. However, since some of the GPS signals will be obstructed in areas with high-rise buildings, such as downtown areas, the vehicle's position may be obtained with the help of conventional systems. As such, the performance of the integrated system is indeed better than either positioning system alone.

The integrated positioning system not only helps the transit authorities to locate their fleet of buses on a digital base map in real time, but also helps in performing other advanced functions (see Figure 10.13). For example, if the bus locations are available in real time, the bus arrival times at the bus stops can be computed reliably, thus minimizing the waiting time at the bus stops. This is a very important feature, especially under severe weather conditions. In addition, the availability of the real-time bus location information enables the transit authorities to dynamically design more efficient bus scheduling, thus improving bus efficiency and customer service. This

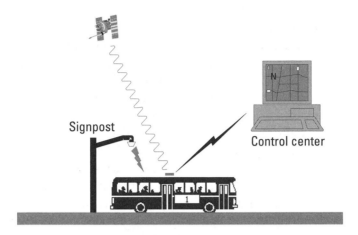

Figure 10.13 GPS for transit systems.

information can be accessed through the Internet, greatly enhancing customer satisfaction [22].

10.13   GPS for the Retail Industry

In today's competitive market, efficiency and cost reduction play a significant role in keeping a retailer in business. GPS integrated with a GIS can help in achieving this goal. Other technologies, such as wireless data communication and speech recognition, are also becoming key components. This section describes how the different technologies are integrated to ensure efficiency and cost reduction in two components of the retail business: delivery and real-time inventory monitoring.

The integrated system consists of two main components: an efficient route analysis for the delivery area and a GPS-based real-time fleet-monitoring system. In designing an efficient route analysis, it is necessary that an up-to-date digital map (street map database) for the delivery area is available. In addition, a retail stores' location database is required. Based on these two databases, as well as the traffic conditions, it is possible to optimize the vehicles' routing and scheduling. A software package such as ESRI's Arclogistics may be used for the purpose of optimization [24]. An optimized real-time fleet-monitoring system requires the availability of both GPS and a suitable wireless data communication. Speech recognition technology may also be used to speed up the delivery service.

To ensure that the GPS-derived position and the map database for the proposed delivery area are compatible, the digital map should be based on WGS 84, the parent datum for GPS. In addition, vector maps rather than raster maps should be used to increase flexibility. Information about driving restrictions, such as one-way streets or vehicle-type restrictions, should accompany the street map database. Retail stores' location is another database that should also be available. Similar to the street map database, the stores' location database should be based on WGS 84. Ultimately, the actual network drive times as a function of the time of day should be considered to optimize the route analysis. Other factors that should be considered in the route analysis include order characteristics, delivery time windows, and the varying vehicles' volume, weight, and operating costs. A good software package should have the capability to use this information to find the best route and stop sequence of each vehicle in the fleet. In addition, the software should have the capability to reroute a vehicle in such cases as road closure or unexpected heavy traffic due to, for example, accidents. Moreover, the software should have the capability to produce detailed maps with directions to the drivers and a summary report to the dispatching center.

With this optimization process, the routing time would be reduced, which means that the number of vehicles needed for the same service is reduced. This ensures that the delivery cost is reduced. Moreover, as a result of the optimization process, it is expected that the delivery time windows will be reduced as well. This means that goods are delivered more efficiently.

The second main component of the integrated system is a GPS-based real-time fleet-monitoring system. Low-cost standalone GPS receivers may be used, as only low-accuracy positioning information is needed. To avoid GPS signal obstruction, it may be necessary for some vehicles in the fleet that operate in urban areas to be equipped with another, complementary, navigation system such as the DR system [22, 25]. In order for the dispatcher to locate a vehicle in the fleet, it is necessary that the GPS position is sent to the dispatching center via a communication link [26]. This could be done in either real time or near real time, depending on the need. In fact, a combination of both might be an ideal solution, if the system is used within a state and outside. At each delivery site, the driver may send some information about the products being delivered. Speech-recognition technology could be used to transfer the driver's voice into digital data that would be sent along with the GPS positioning information. Alternatively, bar codes may be used for the same purpose. The vehicle position should be sent automatically at each delivery site. This may be done by comparing the vehicle's location and the stored location of the delivery site.

The received vehicles' positions along with other information would be overlaid on a base map, which helps the dispatcher to locate each vehicle in the fleet and to know of its contents. This also helps in making efficient decisions, for example, in case of an emergency. Moreover, better estimates for arrival times, based on the actual vehicles' location, can be made. If necessary, dynamic rerouting may be done upon receiving new information on the fleet.

Monitoring the fleet as indicated ensures that each driver follows the preassigned route (i.e., it gives the authorities a means for driver accountability). Although this might be a good feature in the system, the driver's privacy should be considered by, for example, restricting the accessibility of certain information to authorized users.

10.14 GPS for Cadastral Surveying

Cadastral surveys establish property corners, boundaries, and areas of land parcels [1]. Conventional surveying methods have been used, and are still being used, for this purpose. Conventional methods, however, have the drawback that extensive traversing is required. Moreover, extensive clear-cutting and intervening private properties might be required as well. GPS overcomes these conventional-method drawbacks.

Any of the GPS surveying methods, such as kinematic GPS or RTK GPS, can be used depending on the project requirements, location, and other factors. However, the RTK surveying seems to be the most suitable method, especially in unobstructed areas. This is mainly because of its ease of use and the availability of the results while in the field. Inaccessible locations or obstructed areas can be surveyed with integrated systems such as GPS/LRF or GPS/total station.

There are several advantages of using GPS for cadastral surveying. The most important one is that intervisibility between the points is not required with GPS. This means that extensive traversing is eliminated, clear-cutting is not required, and intervening private properties is avoided. Other advantages include the fact that GPS provides user-defined coordinates in a digital format, which can be easily exported to any GIS system for further analysis. The accuracy obtained with GPS is consistent over the entire network; such accuracy is lacked by conventional surveying methods. Also, with GPS, one reference station can support an unlimited number of rover receivers. A number of governmental and private organizations have reported that the use of GPS in cadastral surveying is cost-effective.

10.15   Waypoint Navigation (GPS Stakeout)

Waypoint navigation, or *stakeout*, as it is called by surveyors, provides guidance to a GPS user in reaching his or her destination in the best way (shortest
time and/or distance). By feeding the GPS receiver (or the GPS receiver controller) with the coordinates of his or her destination, a GPS user receives
on-screen guidance instantaneously (see Figure 10.14 for details). Surveyors
use this principle to lay out points and lines.

The idea behind GPS waypoint navigation is simple. As a first step, the
user must feed the GPS receiver (or the GPS controller) with the coordinates
of his or her destination. Most GPS receivers are capable of storing a number
of destination points (waypoints) in their internal memory. The second step
is to let the GPS receiver compute its own position (i.e., the user's position).
Based on the receiver and the destination positions, the built-in receiver
computer calculates the distance and the azimuth of the line connecting the
receiver's position and the destination points. The built-in computer uses the
position information to calculate other parameters, such as the expected
arrival time to the user's destination based on the user's speed. In addition,
the offset distance from the receiver position to the original line between the

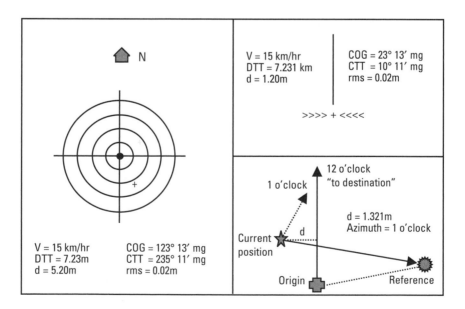

Figure 10.14 GPS waypoint navigation.

starting point and the destination can be calculated. All of this information and other data are displayed on a continuous basis to guide the GPS user.

This guidance information can be displayed in different ways [25]. One of these displays is the bull's-eye, where the destination point is located at the center of the displayed concentric circles while the user's location is displayed as a moving cursor. The top point of the bull's-eye is normally selected to represent the north. The user will reach his or her destination point when the moving cursor stays at the center of the concentric circles. In addition to this, a number of navigation parameters are displayed to help the user.

References

[1] Elfick, M., et al., *Elementary Surveying*, 8th ed., New York: HarperCollins, 1994.

[2] Laser Technology, Inc., "Survey Laser for Forestry," presentation, http://www.lasertech.com/ downloads.html.

[3] Phillips, B., "GPS Field Applications in Forestry Consulting," Global Positioning System in Forestry Workshop, Kelowna, British Columbia, Canada, November 25–28, 1996.

[4] Bauer, W. D., and M. Schefcik, "Using Differential GPS to Improve Crop Yields," *GPS World*, Vol. 5, No. 2, February 1994, pp. 38–41.

[5] Petersen, C., "Precision GPS Navigation for Improving Agriculture Productivity," *GPS World*, Vol. 2, No. 1, January 1991, pp. 38–44.

[6] Smith, B. S., "GPS Grade Control for Construction," *Proc. ION GPS 2000*, 13th Intl. Technical Meeting, Satellite Division, Institute of Navigation, Salt Lake City, UT, September 19–22, 2000, pp. 1034–1037.

[7] El-Rabbany, A., A. Chrzanowski, and M. Santos, "GPS Applications in Civil Engineering," *Ontario Land Surveyor Quarterly*, Summer 2001, pp. 6–8.

[8] Flinn, J. A., C. Waddell, and M. A. Lowery, "Practical Aspects of GPS Implementation at the Morenci Copper Mine," *Proc. ION GPS-99*, 12th Intl. Technical Meeting, Satellite Division, Institute of Navigation, Nashville, TN, September 14–17, 1999, pp. 915–919.

[9] Flinn, J. A., and S. M. Shields, "Optimization of GPS on Track-Dozers at a Large Mining Operation," *Proc. ION GPS-99*, 12th Intl. Technical Meeting, Satellite Division, Institute of Navigation, Nashville, TN, September 14–17, 1999, pp. 927–931.

[10] Dai, L., et al., "GPS and Pseudolite Integration for Deformation Monitoring Applications," *Proc. ION GPS 2000*, 13th Intl. Technical Meeting, Satellite Division, Institute of Navigation, Salt Lake City, UT, September 19–22, 2000, pp. 1–8.

[11] Shields, S., J. Flinn, and A. Obregon, "GPS in the Pits: Differential GPS Application at the Morenci Copper Mine," *GPS World*, Vol. 11, No. 10, October 2000, pp. 34–39.

[12] Jensen, M. H., "Quality Control for Differential GPS in Offshore Oil and Gas Exploration," *GPS World*, Vol. 3, No. 8, September 1992, pp. 36–48.

[13] McLintock, D., G. Deren, and E. Krakiwsky, "Environment Sensitive: DGPS and Barometry for Seismic Surveys," *GPS World*, Vol. 5, No. 2, February 1994, pp. 20–26.

[14] Flood, M., B. Gutelius, and M. Orr, "Airborne Terrain Mapping," *Earth Observation Magazine*, Vol. 6, No. 2, February 1997, pp. 40–42.

[15] Cramer, M., "On the Use of Direct Georeferencing in Airborne Photogrammetry," *Proc. 3rd Intl. Symp. Mobile Mapping Technology*, Cairo, Egypt, January 3–5, 2001, CD-ROM.

[16] Abwerzger, G., "Georeferencing of Laser Scanner Data Using GPS Attitude and Position Determination," *Proc. 3rd Intl. Symp. Mobile Mapping Technology*, Cairo, Egypt, January 3–5, 2001, CD-ROM.

[17] Favey, E., et al., "3D-Laser Mapping and Its Application in Volume Change Detection of Glaciers," *Proc. 3rd Intl. Symp. Mobile Mapping Technology*, Cairo, Egypt, January 3–5, 2001, CD-ROM.

[18] Bowditch, N., *American Practical Navigator*, Bethesda, MD: Defense Mapping Agency Hydrographic/Topographic Center, 1995.

[19] Maxfield, H. E., "Recent Developments in Seafloor Mapping Capabilities," *Hydro International*, Vol. 2, No. 1, January/February 1998, pp. 45–47.

[20] Zhao, Y., *Vehicle Location and Navigation Systems*, Norwood, MA: Artech House, 1997.

[21] Hada, H., et al., "The Internet, Cars, and DGPS: Bringing Mobile Sensors and Global Correction Services On Line," *GPS World*, Vol. 11, No. 5, May 2000, pp. 38–43.

[22] El-Rabbany, A., A. Shalaby, and S. Zolfaghari, "Real-Time Bus Location, Passenger Information and Scheduling for Public Transportation," Presented at GPS meeting, GOEIDE 2000 Conference, Calgary, Alberta, Canada, May 24–26, 2000.

[23] Drane, C., and C. Rizos, *Positioning Systems in Intelligent Transportation Systems*, Norwood, MA: Artech House, 1998.

[24] ESRI, "ArcLogistics Route," *Business Solution News*, http://www.esri.com/software/arclogistics/index.htm.

[25] SOKKIA Corporation, GSR2300 Operation Manual, 1996.

[26] Beerens, J. A. J., "Fleet Monitoring with GPS and Satellite Communications," *GPS World*, Vol. 4, No. 4, April 1993, pp. 42–46.

11

Other Satellite Navigation Systems

11.1 The GLONASS System

GLONASS is a Russian-developed all-weather satellite-based navigation system that has much in common with GPS. The GLONASS program was initiated in the mid 1970s, with the first satellite launched in October 1982. Similar to GPS, GLONASS is a dual-use system that can be accessed by both military and civilian users. The nominal constellation consists of 21 satellites plus 3 active spares at a nominal altitude of 19,100 km [1]. The satellites are evenly distributed over three orbital planes, each containing 8 satellites (nominally spaced evenly within the orbital plane—45 degrees apart) (see Figure 11.1). A GLONASS satellite orbit is nearly circular, with an orbital period of 11 hours and 15 minutes and an inclination of 64.8 degrees to the equator [2–4].

The GLONASS system consists of three segments: space, control, and user. The space segment consists of the nominal 24-satellite constellation. Similar to GPS, each GLONASS satellite transmits a signal that has a number of components: two L-band carriers—C/A-code on L1 (and L2 for the newer generation, GLONASS-M satellites), P-code on both L1 and L2—and a navigation message. However, unlike GPS, each satellite transmits its own carrier frequencies, which depend on the frequency channel number (an integer number assigned to each satellite). During the period 1998 to 2005, GLONASS satellites used the frequency channels 0 to 12. After 2005, the satellites are to use the frequency channels –7 to 6 [2].

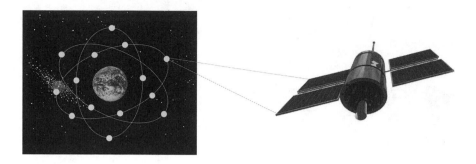

Figure 11.1 The GLONASS orbital configuration. (*Satellite image courtesy of:* Magellan Corporation.)

Table 11.1 shows the corresponding carrier frequencies (note that channels 0 and 12 are used for technical purposes).

Table 11.1

GLONASS Channel Numbers and Corresponding L1 and L2 Frequencies

Frequency Channel Number	Nominal L1 Frequency (MHz)	Nominal L2 Frequency (MHz)
13	1609.3125	1251.6875
12	1608.75	1251.25
11	1608.1875	1250.8125
10	1607.625	1250.375
09	1607.0625	1249.9375
08	1606.5	1249.5
07	1605.9375	1249.0625
06	1605.375	1248.625
05	1604.8125	1248.1875
04	1604.25	1247.75
03	1603.6875	1247.3125
02	1603.125	1246.875
01	1602.5625	1246.4375
00	1602.0	1246.0
−01	1601.4375	1245.5625

Table 11.1 (continued)

Frequency Channel Number	Nominal L1 Frequency (MHz)	Nominal L2 Frequency (MHz)
−02	1600.8750	1245.1250
−03	1600.3125	1244.6875
−04	1599.7500	1244.2500
−05	1599.1875	1243.8125
−06	1598.6250	1243.3750
−07	1598.0625	1242.9375

Source: [2].

The reason for the frequency band shift is to avoid interference with radio astronomers and operators of LEO satellites. As only 12 channels are used, each pair of satellites is assigned the same L1 and L2 frequencies. For operational reasons, the satellite pairs are placed on the opposite sides of the Earth (antipodal), so that a user is not able to track them simultaneously. GLONASS codes are the same for all the satellites. As such, GLONASS receivers use the frequency channel rather than the code to distinguish the satellites. The chipping rates for the P-code and the C/A-code are 5.11 and 0.511 Mbps, respectively. Only the C/A-code is made available to civilians. The GLONASS navigation message is a 50-bps data stream, which provides, among other things, the satellite ephemeris and channel allocation [4]. The GLONASS signal is controlled by one of three onboard cesium clocks. It is worth mentioning that the signal characteristics of the GLONASS system require more complex receiver front-end design (to process the multiple frequencies) than GPS. As a consequence, GLONASS receivers are expected to be larger and more expensive than GPS receivers. However, the GLONASS signal is more resistant to interference than the GPS signal.

The GLONASS system achieved a complete constellation of 24 operating satellites in January 1996. Unfortunately, the number of satellites dropped to seven by May 2001 [5]. Nevertheless, a number of recent launches have taken place, increasing the number of satellites in service to 16 (see Table 11.2). The current constellation includes four new generation satellites known as GLONASS-M satellites ("M" stands for modified). As indicated earlier, GLONASS-M satellites transmit the C/A-code on both L1 and L2, which enables users to correct for the ionospheric delay. In addition, a GLONASS-M satellite has a lifetime of 7 years and improved onboard atomic clocks [1]. Four more GLONASS-M satellites are to be built by the end of 2006 [6]. The Russian government announced a plan to have 18 operational satellites by 2008 [7]. The next generation of GLONASS satellites, known as GLONASS-K, will be an entirely new model based on a nonpressurized platform. Furthermore, the new spacecraft will be smaller and lighter than previous models. A third civilian frequency will be added to GLONASS-K, while its design life will be increased to 10–12 years. Testing of the GLONASS-K satellite is scheduled to start in 2007 [7].

The GLONASS control segment consists of a system control center and a network of command and tracking stations located within Russia. Some of the functions of the control segment include system monitoring, ephemeris estimation, and uploading of navigation data [2]. The user segment includes all military and civilian users suitably equipped with GLONASS receivers. The specification for the expected GLONASS autonomous positioning accuracy for civilian users is 100m and 150m (95 percent) for the horizontal and vertical components, respectively. However, tests in mid-1990s showed that the actual GLONASS autonomous positioning accuracy for civilian users was 26m and 45m (95 percent) for the horizontal and vertical components, respectively [1].

Table 11.2
GLONASS Constellation Status as of April 7, 2006

GLONASS Number	Cosmos Number	Plane/ Slot	Frequency Channel	Launch Date (dd.mm.yy)	Introduction Date (dd.mm.yy)	Status
796	2411	1/01	07	26.12.04	06.02.05	Operating
794	2402	1/02	01	10.12.03	02.02.04	Operating
789	2381	1/03	12	01.12.01	04.01.02	Operating
795	2403	1/04	06	10.12.03	30.01.04	Operating
711	2382	1/05	07	01.12.01	15.04.03	Operating
701	2404	1/06	01	10.12.03	09.12.04	Operating
712	2413	1/07	04	26.12.04	22.12.05	Operating
797	2412	1/08	06	26.12.04	06.02.05	Operating
787	2375	3/17	05	13.10.00	04.11.00	Operating
783	2374	3/18	10	13.10.00	05.01.01	Unusable
798	2417	3/19	03	26.12.05	22.01.06	Operating
793	2396	3/20	11	25.12.02	31.01.03	Operating
792	2395	3/21	05	25.12.02	31.01.03	Operating
791	2394	3/22	10	25.12.02	10.02.03	Operating
714	2419	3/23	–	25.12.05	–	–
713	2418	3/24	–	25.12.05	–	–

Source: http://www.glonass-center.ru

GPS and GLONASS may be integrated at the user level to improve geometry and positioning accuracy, particularly under poor satellite visibility, such as in urban areas. However, there are two problems with GPS/GLONASS integration. The first one is that both systems use different coordinate frames to express the position of their satellites. GPS uses the WGS 84 system, while GLONASS uses the Earth Parameter System 1990 (PZ-90) system. The two systems differ by as much as 20m on the Earth's surface. The transformation parameters between the two systems may be obtained by simultaneously observing reference points in both systems. Various research groups have developed sets of transformation parameters [4, 8]. The second problem with integration is that both systems use different reference times. The offset between the two time systems changes slowly and reaches several tens of microseconds. One way of determining the time offset is by treating it as an additional variable in the receiver solution. Alternatively, a user may obtain the time offset, with a 30-ns precision, through the GLONASS navigation message [2].

11.2 Galileo—The European Global Satellite Navigation System

Galileo is a satellite-based global-navigation system currently being developed in Europe. Galileo is a civil-controlled satellite system to be delivered through a public-private partnership. The system is designed to provide five different service levels as follows [9]:

1. *Open service (OS)*: an open-access service, which provides positioning, velocity, and timing information free of direct charge. The expected positioning accuracy (95 percent) of OS is in the order of 15m and 35m for the horizontal and vertical components, respectively, when single-frequency receivers are used. If dual-frequency data is available, the expected positioning accuracy can be improved to 4m and 8m for the horizontal and vertical components, respectively. The expected timing accuracy (95 percent) is on the order of 30 ns.

2. *Commercial service (CS)*: a restricted-access service, which provides value-added data on dedicated commercial service signal. The expected positioning accuracy (95 percent) of CS is less than 1m when dual-frequency data is used.

3. *Safety of life (SOL)*: an open-access global service, which is intended for safety-critical users (e.g. maritime, aviation, and rail

transportation). With this service, the user will be assured that the received signal is the actual Galileo one. The expected positioning accuracy (95 percent) of SOL is on the order of 4–6m when dual-frequency data is used.

4. *Public regulated service (PRS)*: a restricted-access service, which is available to government-authorized users only. It is used when a higher level of protection is required. The expected positioning accuracy (95 percent) of PRS is on the order of 6.5m and 12m for the horizontal and vertical components, respectively, when dual-frequency data are used.

5. *Support to search and rescue (SAR)*: The SAR service is through a SAR transponder as discussed later.

Three different constellation types were investigated to ensure the optimum selection of the Galileo architecture, namely LEO, MEO, and IGSO (see Chapter 3). Combinations of various constellation types were also studied. Following this study, the Galileo decision makers adopted a constellation of 30 MEO satellites (27 operational plus 3 nonoperational spare satellites). The operational satellites will be evenly distributed over three orbital planes, each containing nine satellites (nominally evenly distributed within a plan—40 degrees apart). Furthermore, each of the orbital planes will contain a nonoperational spare satellite that can be used in the case of a satellite failure [10]. Galileo satellite orbits are nearly circular (nominal eccentricity of about 0.002), with an inclination of about 56 degrees to the equator. The altitude of the Galileo orbit is about 23,222 km above the Earth's surface (i.e., the semimajor axis is about 29,600.318 km). The corresponding orbital period is about 14h 4m 41s, which means that the constellation has a repeat cycle of 17 orbits in 10 days [11]. The selection of the Galileo constellation geometry ensures that more uniform performance is obtained for all regions (i.e., independent of the region's latitude).

Similar to GPS, the Galileo system consists of three main segments: space, ground, and user. The space segment consists of the 30-satellite constellation introduced earlier. Each Galileo satellite carries two payloads: a navigation payload for transmitting navigation data and a search and rescue payload for relaying of alarms from distress beacons to search and rescue organizations [10]. The navigation payload carries two types of atomic clocks, rubidium and passive hydrogen maser, to control the Galileo navigation signal. An operational Galileo satellite will transmit six navigation signals in three different frequency bands: E2-L1-E1 band (1559–1592 MHz), E5 band (1164–1215 MHz), and E6 band (1260–1300 MHz). The six navigation signals are known as L1F, L1P, E6C, E6P, E5a, and E5b [10]. Of these, three are open-access signals, namely L1F, E5a, and E5b. The E6C is a commercial-access signal, while L1P and E6P are restricted-access signals. Each of the open-access and commercial signals includes a data channel and pilot (or dataless) channel. The Galileo system uses three different modulation schemes known as binary phase shift keying (BPSK), binary offset carrier (BOC) and alternative binary offset carrier (AltBOC) (see [10, 12] for details). Table 11.3 summarizes the Galileo navigation signal parameters. It should be pointed out that the E5a signal can be received and processed separately from the E5b signal (i.e., as if it were transmitted alone [10]).

The ground segment of the Galileo system consists of a worldwide network of tracking (sensor) stations, control centers, and uplink stations. The primary task of the ground segment is tracking the Galileo satellites in order to determine and predict satellite locations, system integrity, and other considerations. This information is then packed and uploaded into the Galileo satellites through the C-band link. The user segment includes all users with appropriate receivers to access the various types of Galileo services.

Galileo and GPS have a partial frequency overlap (i.e., the Galileo E2-L1-E1 and E5a will be broadcast using identical frequencies to GPS L1 and L5, respectively). This will make possible Galileo/GPS interoperability and will simplify the receiver front-end design [10]. Galileo will use the Galileo terrestrial reference frame (GTRF), which will be a realization of the international terrestrial reference system (ITRS) (see Appendix A). The difference between GTRF and WGS 84 is expected to be at the few centimeter level. Such a difference can be neglected by most applications. Galileo will also use the Galileo system time (GST), which will be steered to TAI (see Chapter 2). The offset between the GPS system time and GST can be obtained, for example, from traditional time-transfer techniques [10].

The first Galileo demonstrator, known as Galileo in-orbit validation element, or Giove-A, was successfully launched on December 28, 2005 (Figure 11.2). Giove-A transmitted the first Galileo signal on January 12, 2006 [13]. The mission of Giove-A is to secure the use of the allocated Galileo frequencies and demonstrate key technologies for the navigation payloads, among others. A second demonstrator satellite, Giove-B, is scheduled for launch in September 2006. The target date for the Galileo operational service is 2010 or shortly thereafter.

Table 11.3
Parameters of Galileo Navigation Signal

Frequency Band (MHz)	Signal (Channel)	Modulation/ Chipping Rate	PRN Code Encryption	Notes/ Supported Services
E2-L1-E1 (1559–1592)	L1F (data) L1F (pilot)	BOC/1.023 Mbps	Unencrypted ranging code (both L1F data and pilot channels) and navigation data (L1F data channel)	L1F data stream also contains integrity messages and encrypted commercial data; supports OS, CS, and SOL
	L1P	BOC/2.5575 Mbps	Restricted PRS	Supports PRS
E6 (1260–1300)	E6C (data) E6C (pilot)	BPSK/5.115 Mbps	Restricted commercial	Supports commercial service
	E6P	BOC/5.115 Mbps	Restricted PRS	Supports PRS
E5 (1164–1215)	E5a (data) E51 (pilot) E5b (data) E5b (pilot)	AltBOC/10.23 Mbps	Unencrypted ranging code (both L1F data and pilot channels) and navigation data (L1F data channel)	Supports OS Notes and supported services are similar to L1F

11.3 Chinese Regional Satellite Navigation System (Beidou System)

The Beidou system is a regional, two-dimensional, all-weather positioning and navigation system developed by China. The first generation of the Beidou system consists of three satellites (two operational and one backup) placed in geosynchronous orbits [14]. The two operational satellites were launched in October and December 2000, respectively, while the backup satellite was launched in May 2003 [14]. The Beidou system provides navigation and communication services to both Chinese military and civilian users, and can be used to augment the GPS and GLONASS systems [1]. The system provides coverage for the region between latitudes of 5 degrees to 55 degrees N and longitudes of 70 degrees to 140 degrees E. The Beidou system

Figure 11.2 Artist's impression of GIOVE-A in orbit. (*Source:* http://www.esa.int/ SPECIALS/Galileo_Launch/SEMQ36MZCIE_1.html.)

has a ground segment consisting of a central control station and ground correction stations. The user segment consists of a receiving/transmitting terminal.

Unlike GPS, GLONASS, and Galileo, which are one-way-ranging systems, Beidou is a two-way ranging system [15]. That is, the central station sends an inquiry signal to the users through a Beidou satellite. Upon receiving the inquiry signal, the users respond by transmitting a signal to the two Beidou satellites, which will then be transmitted to the central station (Figure 11.3). Upon receiving the responding signals (from the two satellites), the central station estimates the signal travel time, which can be used, along with the known Beidou satellite locations and an estimate of the user's altitude, to estimate the user's location. Once estimated, the central station transmits the positioning data to the user through the Beidou satellite [14]. The expected horizontal positioning accuracy of the Beidou system is at the level of 20m to 100m, depending on whether or not the ground correction stations are used. It should be pointed out that the satellites broadcast navigation signals to users at 2,492 MHz (S-band), while the user terminal transmits a short-burst, spread-spectrum response at 1,616 MHz (L-band) [1].

China is planning to build its second generation satellite positioning and navigation system, known as Beidou-2, which will have more satellites and more coverage area [15].

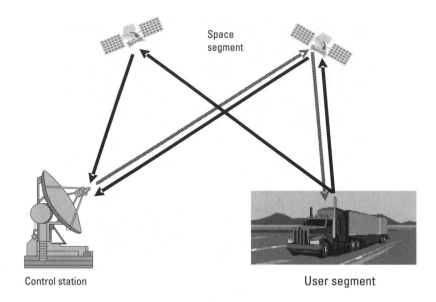

Space
segment

Control station   User segment

Figure 11.3 Operational concept of Beidou system.

11.4   The Japanese QZSS Satellite Navigation System

To meet commercial demands, the government of Japan, in collaboration
with Japanese industry, is developing a regional satellite navigation system
known as the Quasi-Zenith Satellite System (QZSS). The Japanese QZSS
system is an autonomous, yet GPS-compatible, satellite-based navigation system to serve Japan as well as all of Asia [16]. This means that the QZSS system will provide independent satellite-based positioning, velocity, and
timing services, while taking advantage of the availability of the GPS signals
and industry. In addition, the QZSS system is to provide broadcast and communication services (in S-band) to Japan.

The constellation of the QZSS system will consist of three highly
inclined geosynchronous orbits, each containing one satellite. The orbital
inclination will be 45 degrees, while the orbital eccentricity will be 0.099.
The satellite altitudes will be 39,970 km and 31,602 km at apogee and perigee, respectively. The constellation is designed so that one of the three satellites will always be over Japan [17]. The first QZSS satellite is expected to be
launched in 2008, while the other two are expected to be launched in 2009.

To ensure interoperability with GPS, the QZSS system will use the
same GPS L1, L2, and L5 signals. In addition, the system will use a new

experimental signal, known as LEX, at 1278.75 MHz (it shares the frequency band with Galileo E6 signal) with higher data rate message. Similar to Galileo, QZSS will use the ITRS as the geodetic reference frame. However, the QZSS time frame will be offset from TAI by 19 seconds (i.e., equivalent to that of GPS) [17].

The future generation of the Japanese regional navigation satellite system will contain seven quasi-zenith and geostationary satellites. This would allow for four satellites to be visible at relatively high elevation angles [16].

References

[1] Feairheller, S. and R. Clark, "Other Satellite Navigation Systems," in *Understanding GPS: Principles and Applications*, 2nd Ed., E. D. Kaplan and C. J. Hegarty (Eds.), Norwood, MA: Artech House, 2006.

[2] Global Navigation Satellite System—Glonass, Interface Control Document, Version 5.0, 2002, http://www.glonass-center.ru/icd02_e.zip.

[3] Kleusberg, A., "Comparing GPS and Glonass," *GPS World*, Vol. 1, No. 6, November/December 1990, pp. 52–54.

[4] Langley, R. B., "Glonass: Review and Update," *GPS World*, Vol. 8, No. 7, July 1997, pp. 46–51.

[5] "Navtech Seminars and GPS Supply," *GPS/GNSS* newsletter, May 17, 2001.

[6] Russian News and Information Agency. "Four More Glonass-M Satellites To Be Constructed in 2006," March 20, 2006, http://en.rian.ru/russia/20060320/44549579.html.

[7] "GLONASS To Have 18 Satellites In Orbit In 2008," *Space Daily*, January 19, 2006, http://www.spacedaily.com/reports/GLONASS_To_Have_18_Satellites_In_Orbit_In_2008.html.

[8] Bazlof, Y. A., et al., "Glonass to GPS: A New Coordinate Transformation," *GPS World*, Vol. 10, No. 1, January 1999, pp. 54–58.

[9] Prasad, R., and M. Ruggieri, *Applied Satellite Navigation Using GPS, Galileo, and Augmentation Systems*, Norwood, MA: Artech House, 2005.

[10] Falcone, M., P. Erhard, and G. Hein, "Galileo," in *Understanding GPS: Principles and Applications*, 2nd Ed., E. D. Kaplan and C. J. Hegarty (Eds.), Norwood, MA: Artech House, 2006.

[11] Píriz, R., B. Martín-Peiró, and M. Romay-Merino, "The Galileo Constellation Design: A Systematic Approach," *ION GNSS 18th International Technical Meeting of the Satellite Division*, Long Beach, CA, September 13–16, 2005.

[12] Ward, P. W., "GPS Satellite Signal Characteristics," in *Understanding GPS: Principles and Applications*, 2nd Ed., E. D. Kaplan and C. J. Hegarty (Eds.), Norwood, MA: Artech House, 2006.

[13] European Space Agency, "GIOVE A transmits loud and clear," http://www.esa.int/esaCP/SEM21VMVGJE_index_1.html#subhead2.

[14] "Beidou-1 Satellite Navigation System," http://www.sinodefence.com/space/satellite/beidou1.asp.

[15] Kaplan, E. D., "Introduction," in *Understanding GPS: Principles and Applications*, 2nd Ed., E. D. Kaplan and C. J. Hegarty (Eds.), Norwood, MA: Artech House, 2006.

[16] Takahashi, H., "Japanese Regional Navigation Satellite System—the JRANS Concept," *ION GNSS 18th International Technical Meeting of the Satellite Division*, Long Beach, CA, September 13–16, 2005.

[17] Maeda, H., "QZSS Overview and Interoperability," *ION GNSS 18th International Technical Meeting of the Satellite Division*, Long Beach, CA, September 13–16, 2005.

Appendix A:
Geodetic Principles—Datums, Coordinate Systems, and Map Projections

The ability of GPS to determine the precise location of a user anywhere, under any weather conditions, attracted millions of users worldwide from various fields and backgrounds. With advances in GPS and computer technologies, GPS manufacturers were able to come up with very user-friendly systems. However, one common problem that many newcomers to the GPS field face is the issue of datums and coordinate systems, which require some geodetic background. This chapter tackles the problem of datums and coordinate systems in detail. As in the previous chapters, complex mathematical formulas are avoided. As many users are interested in the horizontal component of the GPS position, the issue of map projections is also introduced. For the sake of completeness, the height systems are introduced as well, at the end of this chapter.

A.1   What Is a Datum?

The fact that the topographic surface of the Earth is highly irregular makes it difficult for geodetic calculations (e.g., the determination of the user's location) to be performed. To overcome this problem, geodesists adopted a smooth mathematical surface, called the reference surface, to approximate the irregular shape of the Earth (more precisely to approximate the global mean sea level, the geoid) [1, 2]. One such mathematical surface is the sphere, which has been widely used for low-accuracy positioning. For high-accuracy positioning, such as GPS positioning, however, the best mathematical surface to approximate the Earth and at the same time keep the calculations as simple as possible was found to be the biaxial ellipsoid (see Figure A.1). The biaxial reference ellipsoid, or simply the reference ellipsoid, is obtained by rotating an ellipse around its minor axis, b [2]. Similar to the ellipse, the biaxial reference ellipsoid can be defined by the semiminor and semimajor axes (a, b) or the semimajor axis and the flattening (a, f), where $f = 1-(b / a)$.

An appropriately positioned reference ellipsoid is known as the geodetic datum [2]. In other words, a geodetic datum is a mathematical surface, or a reference ellipsoid, with a well-defined origin (center) and orientation. For example, a geocentric geodetic datum is a geodetic datum with its origin coinciding with the center of the Earth. It is clear that there is an infinite number of geocentric geodetic datums with different orientations. Therefore, a geodetic datum is uniquely determined by specifying eight parameters: two parameters to define the dimension of the reference ellipsoid, three parameters to define the position of the origin, and three parameters to define the orientation of the three axes with respect to the Earth. Table A.1 shows

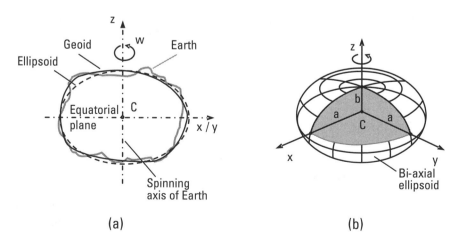

Figure A.1 (a) Relationship between the physical surface of the Earth, the geoid, and the ellipsoid; (b) ellipsoidal parameters.

Table A.1

Examples of Reference Systems and Associated Ellipsoids

Reference System	Ellipsoid	a (m)	$1/f$
WGS 84	WGS 84	6378137.0	298.257223563
NAD 83	GRS 80	6378137.0	298.257222101
NAD 27	Clarke 1866	6378206.4	294.9786982

some examples of three common reference systems and their associated ellipsoids [3].

In addition to the geodetic datum, the so-called vertical datum is used in practice as a reference surface to which the heights (elevations) of points are referred [2]. Because the height of a point directly located on the vertical datum is zero, such a vertical reference surface is commonly known as the surface of zero height. The vertical datum is often selected to be the geoid—the surface that best approximates the mean sea level on a global basis; see Figure A.1(a).

In the past, positions with respect to horizontal and vertical datums have been determined independent of each other [2]. However, with the advent of space geodetic positioning systems such as GPS, it is possible to determine the three-dimensional positions with respect to a three-dimensional reference system.

A.2 Geodetic Coordinate System

A coordinate system is defined as a set of rules for specifying the locations (also called coordinates) of points [4]. This usually involves specifying an origin of the coordinates as well as a set of reference lines (called axes) with known orientation. Figure A.2 shows the case of a three-dimensional coordinate system that uses three reference axes (x, y, and z) that intersect at the origin (C) of the coordinate system.

Coordinate systems may be classified as one-dimensional, two-dimensional, or three-dimensional coordinate systems, according to the number of coordinates required to identify the location of a point. For example, a one-dimensional coordinate system is needed to identify the height of a point above the sea surface.

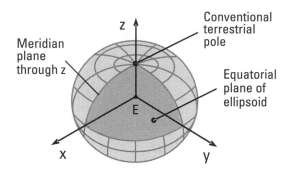

Figure A.2 Three-dimensional coordinate system.

Coordinate systems may also be classified according to the reference surface, the orientation of the axes, and the origin. In the case of a three-dimensional geodetic (also known as geographic) coordinate system, the reference surface is selected to be the ellipsoid. The orientation of the axes and the origin are specified by two planes: the meridian plane through the polar or z-axis (a meridian is a plane that passes through the north and south poles) and the equatorial plane of the ellipsoid (see Figure A.2 for details).

Of particular importance to GPS users is the three-dimensional geodetic coordinate system. In this system, the coordinates of a point are identified by the geodetic latitude (ϕ), the geodetic longitude (λ), and the height above the reference surface (h). Figure A.3 shows these parameters. Geodetic coordinates (ϕ, λ, and h) can be easily transformed to Cartesian coordinates (x, y, and z) as shown in Figure A.3(b) [2]. To do this, the ellipsoidal parameters (a and f) must be known. It is also possible to transform the geodetic

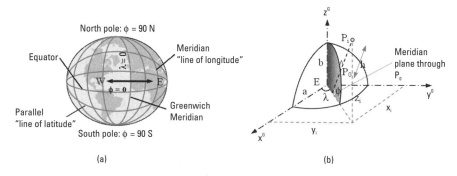

Figure A.3 (a) Concept of geodetic coordinates; (b) geodetic and Cartesian coordinates.

coordinates (ϕ and λ) into a rectangular grid coordinate (e.g., Northing and Easting) for mapping purposes [5].

It is useful for some applications to express the coordinates of a point in a system known as the East-North-Up (ENU) system, which is related to the three-dimensional geodetic coordinate system discussed earlier (see Figure A.4). The ENU has its origin at the user's (i.e., point's) location; therefore, it is also known as the local-level system. The ENU system is a right-handed system with the E-axis (also known as the 1-axis) and N-axis (also known as the 2-axis) pointing toward the geodetic east and north, respectively. The U-axis (also known as the 3-axis) of the ENU system points upward and is perpendicular to the ellipsoidal surface at the user's location (Figure A.4). It is possible to transform the ENU coordinates into geodetic coordinates, and vice versa.

A.2.1 Conventional Terrestrial Reference System

The Conventional Terrestrial Reference System (CTRS) is a three-dimensional geocentric coordinate system (i.e., its origin coincides with the center of the Earth; see Figure A.2). The CTRS is rigidly tied to the Earth (i.e., it rotates with the Earth [5]). It is therefore also known as the Earth-centered, Earth-fixed (ECEF) coordinate system.

The orientation of the axes of the CTRS is defined as follows: The z-axis points toward the conventional terrestrial pole, which is defined as the average location of the pole during the period 1900–1905 [3]. The x-axis is

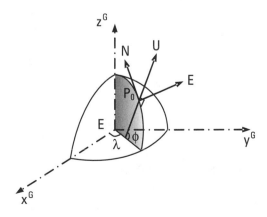

Figure A.4 ENU coordinate system.

defined by the intersection of the terrestrial equatorial plane and the meridianal plane that contains the mean location of the Greenwich observatory (known as the mean Greenwich meridian). It is clear from the definition of the x- and z-axes that the xz-plane contains the mean Greenwich meridian. The y-axis is selected to make the coordinate system right-handed (i.e., 90 degrees east of the x-axis, measured in the equatorial plane). The three axes intersect at the center of the Earth, as shown in Figure A.2.

The CTRS must be positioned with respect to the Earth (known as realization) to be of practical use in positioning [2]. This is done by assigning coordinate values to a selected number of well-distributed reference stations. One of the most important CTRSs is the ITRS, which is realized as the ITRF. The ITRF solution is based on the measurements from globally distributed reference stations using GPS and other space geodetic systems. It is therefore considered to be the most accurate coordinate system [6]. The ITRF is updated every 1 to 3 years to achieve the highest possible accuracy. The most recent version at the time of this writing is the ITRF2000, with a new version, ITRF2005, currently being developed.

A.2.2 The WGS 84 and NAD 83 Systems

The WGS 84 is a three-dimensional, Earth-centered reference system developed by the former U.S. Defense Mapping Agency now incorporated into a new agency, NGA. WGS 84 is the official GPS reference system. In other words, a GPS user who employs the broadcast ephemeris in the solution process will obtain his or her coordinates in the WGS 84 system. The WGS 84 utilizes the CTRS combined with a reference ellipsoid that is identical, up to a very slight difference in flattening, with the ellipsoid of the Geodetic Reference System of 1980 (GRS 80); see Table A.1. The latter was recommended by the International Association of Geodesy for use in geodetic applications [5]. WGS 84 was originally established (realized) using a number of Doppler stations. It was then updated several times to bring it as close as possible to the ITRF reference system. With the most recent update, WGS 84 is coincident with the ITRF at the centimeter accuracy level [7].

In North America, another nominally geocentric datum, the North American Datum of 1983 (NAD 83), is used as the legal datum for spatial positioning. NAD 83 utilizes the ellipsoid of the GRS 80, which means that the size and shape of both WGS 84 and NAD 83 are almost identical. The original realization of NAD 83 was done in 1986, by adjusting "primarily" classical geodetic observations that connected a network of horizontal control stations spanning North America with several hundred observed Doppler positions. Initially, NAD 83 was designed as an Earth-centered reference system [8]. However, with the development of more accurate techniques, it was found that the origin of NAD 83 is shifted by about 2m from the true Earth's center. In addition, access to NAD 83 was provided mainly through a horizontal control network, which has a limited accuracy due to the accumulation of errors. To overcome these limitations, NAD 83 was tied to ITRF using 12 common, very long baseline interferometry (VLBI) stations located in both Canada and the United States (VLBI is a highly accurate, yet complex, space positioning system). This resulted in an improved realization of the NAD 83, which is referred to as NAD 83 (CSRS) and NAD 83 (NSRS) in both Canada and the United States, respectively [8]. The acronyms CSRS and NSRS refer to the Canadian Spatial Reference System and National Spatial Reference System, respectively. It should be pointed out that, due to the different versions of the ITRF, it is important to define to which epoch the ITRF coordinates refer.

A.3 What Coordinates Are Obtained with GPS?

The satellite coordinates as given in the broadcast ephemeris will refer to the WGS 84 reference system. Therefore, a GPS user who employs the broadcast ephemeris in the estimation process will obtain his or her coordinates in the WGS 84 system as well. However, if a user employs the precise ephemeris obtained from the IGS service (Chapter 7), his or her solution will be referred to the ITRF reference system. Some agencies provide the precise ephemeris in various formats. For example, NRCan provides its precise ephemeris data in both the ITRF and the NAD 83 (CSRS) formats.

The question that may arise is what happens if the available reference (base) station coordinates are in NAD 83 rather than in WGS 84? The answer to this question varies, depending on whether the old or the improved NAD 83 system is used. Although the sizes and shapes of the reference ellipsoids of the WGS 84 and the old NAD 83 are almost identical, their origins are shifted by more than 2m with respect to each other [3]. This shift causes a discrepancy in the absolute coordinates of points when expressed in both reference systems. In other words, a point on the Earth's surface will have WGS 84 coordinates that are different from its coordinates in the old NAD 83. The largest coordinate difference is in the height component (about 0.5m). However, the effect of this shift on the relative GPS positioning is negligible. For example, if a user applies the NAD 83 coordinates for the reference station instead of its WGS 84 coordinates, his or her solution will be in the NAD 83 reference system with a negligible error (typically at the millimeter level).

A.4 Datum Transformations

As stated in Section A.1, in the past, positions with respect to horizontal and vertical datums have been determined independent of each other [2]. In addition, horizontal datums were nongeocentric and were selected to best fit certain regions of the world (Figure A.5). As such, those datums were commonly called local datums. More than 150 local datums have been used by different countries of the world. An example of the local datums is the North American datum of 1927 (NAD 27). With the advent of space geodetic positioning systems such as GPS, it is now possible to determine global three-dimensional geocentric datums.

Old maps were produced with the local datums, while new maps are mostly produced with the geocentric datums. Therefore, to ensure consistency, it is necessary to establish the relationships between the local datums and the geocentric datums, such as WGS 84. Such a relationship is known as the datum transformation (see Figure A.6). NGA has published the transformation parameters between WGS 84 and the various local datums used in many countries. Many GPS manufacturers currently use these parameters within their processing software packages. It should be clear, however, that these transformation parameters are only approximate and should not be used for

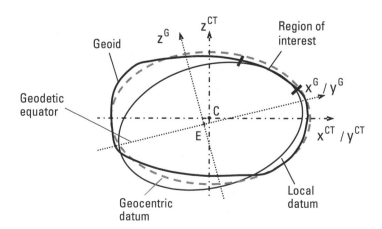

Figure A.5 Geocentric and local datums.

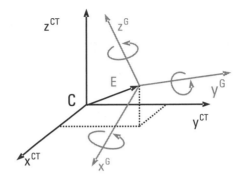

Example: NAD 27 shifts approximately: −9 m, 160 m, 176 m

Figure A.6 Datum transformations.

precise GPS applications. In Toronto, for example, a difference as large as several meters in the horizontal coordinates is obtained when applying NGA's parameters (WGS 84 to NAD 27) as compared with the more precise National Transformation software (NTv2) produced by NRCan. Such a difference could be even larger in other regions. The best way to obtain the transformation parameters is by comparing the coordinates of well-distributed common points in both datums.

A.5 Map Projections

Map projection is defined, from the geometrical point of view, as the transformation of the physical features on the curved Earth's surface onto a flat surface called a map (see Figure A.7). However, it is defined, from the

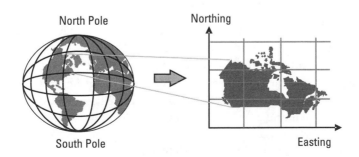

Figure A.7 Concept of map projection.

mathematical point of view, as the transformation of geodetic coordinates (ϕ, λ) obtained from, for example, GPS, into rectangular grid coordinates often called easting and northing. This is known as the direct map projection [2, 4]. The inverse map projection involves the transformation of the grid coordinates into geodetic coordinates. Rectangular grid coordinates are widely used in practice, especially the geomatics-related works. This is mainly because mathematical computations are performed easier on the mapping plane as compared with the reference surface (i.e., the ellipsoid).

Unfortunately, because of the difference between the ellipsoidal shape of the Earth and the flat projection surface, the projected features suffer from distortion [3]. In fact, this is similar to trying to flatten the peel of one-half of an orange; we will have to stretch portions and shrink others, which results in distorting the original shape of the peel. A number of projection types have been developed to minimize map distortions. In most of the GPS applications, the so-called conformal map projection is used [2]. With conformal map projection, the angles on the surface of the ellipsoid are preserved after being projected on the flat projection surface (i.e., the map). However, both the areas and the scales are distorted; remember that areas are either squeezed or stretched [9]. The most popular conformal map projections are transverse Mercator, universal transverse Mercator (UTM), and Lambert conformal conic projections.

It should be pointed out that not only should the projection type accompany the grid coordinates of a point, but the reference system should as well. This is because the geodetic coordinates of a particular point will vary from one reference system to another. For example, a particular point will have different pairs of UTM coordinates if the reference systems are different (e.g., NAD 27 and NAD 83).

A.5.1 Transverse Mercator Projection

Transverse Mercator projection (also known as Gauss-Krüger projection) is a conformal map projection invented by Johann Lambert (Germany) in 1772 [9]. It is based on projecting the points on the ellipsoidal surface mathematically onto an imaginary transverse cylinder (i.e., its axis lies in the equatorial plane). The cylinder can be either a tangent to the ellipsoid along a meridian, called the central meridian, or a secant cylinder (see Figure A.8 for the case of tangent cylinder). In the latter case, two small complex curves at equal distance from the central meridian are produced.

Upon cutting and unfolding the imaginary cylinder, the required flat map (i.e., transverse Mercator projection) is produced. Again, it should be

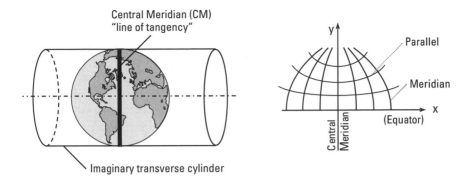

Figure A.8 Transverse Mercator projection.

understood that the transverse cylinder is only an imaginary surface. As explained earlier, the projection is made mathematically through the transformation of the geodetic coordinates into the grid coordinates.

In the case of a tangent cylinder, all features along the line of tangency, the central meridian, are mapped without distortion. This means that the scale, which is a measure of the amount of distortion, is true (equals one) along the central meridian. As we move away from the central meridian, the projected features will suffer from distortion. The farther we are from the central meridian, the greater the distortion will be. In fact, the scale factor increases symmetrically as we move away from the central meridian. This is why this projection is more suitable for areas that are long in the north-south direction.

In the case of a secant cylinder, all features along the two small complex curves will be mapped without distortion (see Figure A.9). The scale is true along the two small complex curves, not the central meridian. Similar to the tangent cylinder case, the distortion increases as we move away from the two small complex curves.

A.5.2 Universal Transverse Mercator

The UTM is a map projection that is based completely on the original transverse Mercator, with a secant cylinder (Figure A.9). With UTM, however, the Earth (i.e., the ellipsoid) is divided into 60 zones of the same size; each zone has its own central meridian that is located at exactly the middle of the zone [9]. This means that each zone covers 6 degrees of longitude, 3 degrees on each side of the zone's central meridian. Each zone is projected separately (i.e., the imaginary cylinder will be rotated around the Earth), which leads to

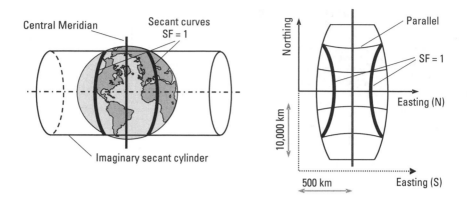

Figure A.9 UTM projection.

a much smaller distortion compared with the original transverse Mercator projection. Each zone is assigned a number ranging from 1 to 60, starting from l = 180 degrees W, and increases eastward (i.e., zone 1 starts at 180 degrees W and ends at 174 degrees W with its central meridian at 177 degrees W); see Figure A.10.

UTM utilizes a scale factor of 0.9996 along the zone's central meridian (Figure A.9). The reason for selecting this scale factor is to have a more uniformly distributed scale, with a minimum deviation from one, over the entire zone. For example, at the equator, the scale factor changes from 0.9996 at the central meridian to 1.00097 at the edge of the zone, while at midlatitude (ϕ = 45 degrees N), the scale changes from 0.9996 at the central meridian to

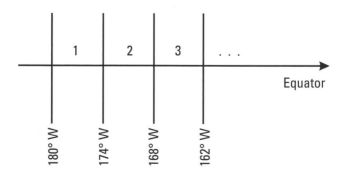

Figure A.10 UTM zoning.

1.00029 at the edge of the zone. This shows how the distortion is kept at a minimal level with UTM.

To avoid negative coordinates, the true origin of the grid coordinates (i.e., where the equator meets the central meridian of the zone) is shifted by introducing the so-called false northing and false easting (Figure A.9). The false northing and false easting take different values, depending on whether we are in the northern or the southern hemisphere. For the northern hemisphere, the false northing and false easting are 0.0 km and 500 km, respectively, while for the southern hemisphere, they are 10,000 km, and 500 km, respectively.

A final point to be made here is that UTM is not suitable for projecting the polar regions. This is mainly due to the many zones to be involved when projecting a small polar area. Other projection types, such as the stereographic double projection, may be used (see Section A.5.5).

A.5.3 Modified Transverse Mercator

The modified transverse Mercator (MTM) projection is another projection that, similar to the UTM, is based completely on the original transverse Mercator, with a secant cylinder [9]. MTM is used in some Canadian provinces, such as the province of Ontario. With MTM, a region is divided into zones of 3 degrees of longitude each (i.e., 1.5 degrees on each side of the zone's central meridian). Similar to UTM, each zone is projected separately, which leads to a small distortion. In Canada, the first zone starts at some point just east of Newfoundland (λ = 51 degrees 30 minutes W) and increases westward. Canada is covered by a total of 32 zones, while the province of Ontario is covered by 10 zones (zones 8 through 17). Figure A.11 shows zone 10, where the city of Toronto is located.

MTM utilizes a scale factor of 0.9999 along the zone's central meridian (Figure A.11). This leads to even less distortion throughout the zone, as compared with the UTM. For example, at a latitude of ϕ = 43.5 degrees N, the scale factor changes from 0.9999 at the central meridian to 1.0000803 at the boundary of the zone. This shows how the scale variation and, consequently, the distortion are minimized with MTM [9]. This, however, has the disadvantage that the number of zones is doubled.

Similar to UTM, to avoid negative coordinates, the true origin of the grid coordinates is shifted by introducing the false northing and false easting. As Canada is completely located in the northern hemisphere, there is only one false northing and one false easting of 0.0m and 304,800m, respectively (see Figure A.11).

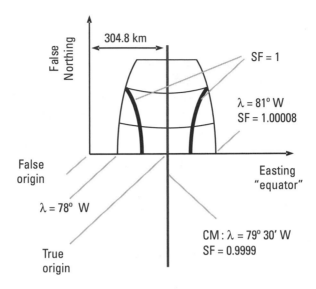

Figure A.11 MTM projection.

A.5.4   Lambert Conical Projection

Lambert conical projection is a conformal map projection developed by Johann Lambert (Germany) in 1772 (the same year in which he developed the transverse Mercator projection). It is based on projecting the points on the ellipsoidal surface mathematically onto an imaginary cone [9]. The cone may either touch the ellipsoid along one of the parallels or intersect the ellipsoid along two parallels. The resulting parallels are called the standard parallels (i.e., one standard parallel is produced in the first case, while two standard parallels are produced in the second case). Upon cutting and unfolding the imaginary cone, the required flat map is produced (Figure A.12).

As with the case of the transverse Mercator projection, all features along the standard parallels are mapped without distortion. As we move away from the standard parallels, the projected features will suffer from distortion. This means that this projection is more suitable for areas that extend in the east-west direction.

This projection is designed so that all the parallels are projected as parts of concentric circles with the center at the apex of the cone, while all the meridians are projected as straight lines converging at the apex of the cone. In other words, the meridians will be the radii of concentric centers. A central

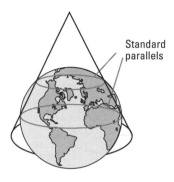
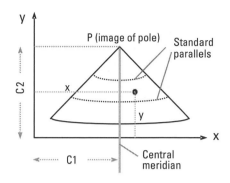

Figure A.12 Lambert conical projection.

meridian that nearly passes through the middle of the area to be mapped is selected to establish the direction of the grid north (i.e., the y-axis). To avoid negative coordinates, the origin of the grid coordinates is shifted by introducing two constants, C1 and C2 (see Figure A.12). The values of C1, C2, and the latitude of the standard parallels are determined by the mapping authorities.

A.5.5 Stereographic Double Projection

The stereographic double projection is a map projection used in some parts of the world, including the Canadian province of New Brunswick. With this mapping projection, points on the reference ellipsoid are projected onto the projection plane through an intermediate surface: an imaginary sphere [9]. In other words, the projection is done in two steps, hence the name *double projection*. First, features on the reference ellipsoid are conformally projected onto an imaginary sphere. Second, features on the sphere are conformally projected onto a tangent or a secant plane to produce the required map (Figure A.13). The latter projection is known as stereographic projection.

There are three cases of stereographic projection to be obtained depending on the position of the projection plane relative to the sphere (i.e., the origin O). If the origin is selected at one of the poles of the sphere, the projection is called polar stereographic. However, if it is selected at some point on the equator of the sphere, the projection is called transverse or equatorial stereographic. The general case in which the origin is selected at an arbitrary point is called oblique stereographic. In the last case, the meridian passing through the map origin is projected as a straight line. All other

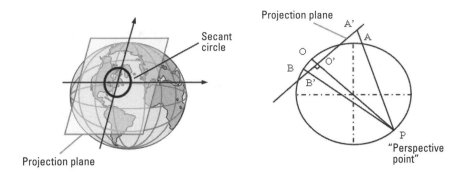

Figure A.13 Stereographic double projection.

meridians and parallels are projected as circles. In New Brunswick, a secant projection plane is used with an origin selected at ϕ = 46 degrees 30 minutes N and λ = 66 degrees 30 minutes W.

In the stereographic projection, a perspective point (P) is first selected to be diametrically opposite to the origin (O). If a secant projection plane is used, a point (A) on the surface of the sphere is projected by drawing a line (PA) and extending it outward to A′ on the projection plane (see Figure A.13). Points inside the secant circle, such as point B, are projected inward. As discussed before, features along the secant circle are projected without distortion, while other features suffer from distortion. In New Brunswick, a scale factor of 0.999912 is selected at the origin. Similar to the previous three map projections, a false northing and a false easting are introduced to avoid negative coordinates.

A.6 Local Arbitrary Mapping Systems

When surveying small areas, it is often more appropriate to employ a user-defined local plane coordinate system. In this case, the curved Earth's surface may be considered a plane surface with a negligible amount of distortion. To establish a local coordinate system with GPS, a set of points with known coordinate values in both the WGS 84 and the local system must be available [5].

By comparing the coordinates of the common points (i.e., points with known coordinates in both the local system and the WGS 84 system), the transformation parameters may be obtained using the least squares technique. These transformation parameters will be used to transform all the new

GPS-derived coordinates into the local coordinate system. It should be noted that the better the distribution of the common points, the better the solution will be (see Figure A.14). The number of common points also plays an important role. The greater the number of common points, the better the solution will be [2].

Establishing a local coordinate system is usually done in either of two ways. One way is to supply the transformation parameters software (usually provided by the manufacturers of the GPS receivers) with the coordinates of the common points in both systems, if they are available. The software will then compute the transformation parameters, which once downloaded into the GPS data collector will be used to automatically transform all the new coordinates into the local coordinate system. Alternatively, if the coordinates of a set of points are known only in the local coordinate system, the user may occupy those points with the rover receiver to obtain their coordinates in the WGS 84 system. RTK GPS surveying (see Chapter 5) is normally used for this purpose. This allows the determination of the transformation parameters while in the field.

A.7 Height Systems

The height (or elevation) of a point is defined as the vertical distance from the vertical datum to the point (Figure A.15). As stated in Section A.1, the

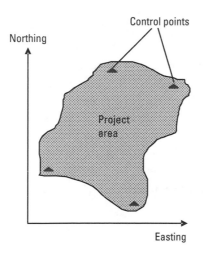

Figure A.14 Local arbitrary mapping system.

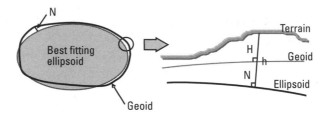

Figure A.15 Height systems.

geoid is often selected to be the vertical datum [2]. The height of a point above the geoid is known as the orthometric height (H). It can be positive or negative, depending on whether the point is located above or below the geoid. Because they are physically meaningful, orthometric heights are often needed in practice and are usually found plotted on topographic maps [2].

In some cases, such as the case of GPS, the obtained heights are referred to the reference ellipsoid, not the geoid (Figure A.15). Therefore, these heights are known as the ellipsoidal heights. An ellipsoidal height (h) can also be positive or negative, depending on whether the point is located above or below the surface of the reference ellipsoid. Unfortunately, ellipsoidal heights are purely geometrical and do not have any physical meaning. As such, the various geomatics instruments (e.g., the total stations) cannot directly sense them.

The geoid-ellipsoid separation (N) is known as the geoidal height or undulation (Figure A.15). This distance can reach up to about 100m, and it can be positive or negative, depending on whether the geoid is above or below the reference ellipsoid at a particular point [10]. Accurate information about the geoidal height leads to the determination of the orthometric height through the ellipsoidal height, and vice versa. Geoid models that describe the geoidal heights for the whole world have been developed. Unfortunately, these models do not have consistent accuracy levels everywhere, mainly because of the lack of local gravity data and the associated height information in some parts of the world [10]. Many GPS receivers and software packages have built-in geoid models for automatic conversion between orthometric and ellipsoidal heights. However, care must be taken when applying them, as they are usually low-accuracy models.

References

[1] Torge, W., *Geodesy*, 3rd ed., New York: Walter de Gruyter, 2001.

[2] Vanicek, P., and E. J. Krakiwsky, *Geodesy: The Concepts*, 2nd ed., New York: North Holland, 1986.

[3] Leick, A., *GPS Satellite Surveying*, 2nd ed., New York: Wiley, 1995.

[4] National Geodetic Survey, "Geodetic Glossary," U.S. Department of Commerce, NOAA, Rockville, MD, 1986.

[5] Hoffmann-Wellenhof, B., H. Lichtenegger, and J. Collins, *Global Positioning System: Theory and Practice*, 3rd ed., New York: Springer-Verlag, 1994.

[6] Boucher, C., and Z. Altamimi, "International Terrestrial Reference Frame," *GPS World*, Vol. 7, No. 9, September 1996, pp. 71–74.

[7] Merrigan, M. J., et al., "A Refinement to the World Geodetic System 1984 Reference Frame," *ION GPS 2002*, Portland, OR, September 24–27, 2002.

[8] Craymer, M., R. Ferland, and R. Snay, "Realization and Unification of NAD83 in Canada and the U.S. Via the ITRF," *Proc. Intl. Symp. Intl. Assoc. of Geodesy*, Sec. 2, "Towards an Integrated Geodetic Observing System (IGGOS)," Munich, Germany, October 5–9, 1998.

[9] Krakiwsky, E. J., "Conformal Map Projections in Geodesy," *L.N. No. 37*, Department of Geodesy and Geomatics Engineering, University of New Brunswick, Fredericton, New Brunswick, Canada, 1973.

[10] Schwarz, K. P., and M. G. Sideris, "Heights and GPS," *GPS World*, Vol. 4, No. 2, February 1993, pp. 50–56.

Appendix B:
GPS Accuracy and Precision Measures

The term *accuracy* is used to express the degree of closeness of a measurement, or the obtained solution, to the true value. The term *precision*, however, is used to describe the degree of closeness of repeated measurements of the same quantity to each other. In the absence of systematic errors, accuracy and precision would be equivalent [1]. For this reason, the two terms are used indiscriminately in many practical purposes. Accuracy can be measured by a statistical quantity called the standard deviation, assuming that the GPS measurements contain no systematic errors or blunders. The lower the standard deviation is, the higher the accuracy.

For the one-dimensional case, for example, when measuring the length of a line between two points, the accuracy is expressed by the so-called root mean square (rms). The rms is associated with a probability level of 68.3 percent. For example, the accuracy of the static GPS surveying could be expressed as 5 mm + 1 ppm (rms). This means that there is a 68.3 percent chance (or probability) that we get an error of less than or equal to 5 mm + 1 mm for every kilometer. In other words, if we measure a 10-km baseline, then there is a 68.3 percent chance that we get an error of less than or equal to 15 mm in the measured line.

Horizontal component (e.g., easting and northing) accuracy, a two-dimensional case, is expressed by either circular error probable (CEP) or twice distance rms (2drms). CEP means that there is a 50 percent chance that the true horizontal position is located inside a circle of radius equal to the value of CEP [1]. The corresponding probability level of the 2drms varies from 95.4 percent to 98.2 percent, depending on the relative values of the errors in the easting and northing components. The ratio of the 2drms to the CEP varies from 2.4 to 3. This means that an accuracy of 40m (CEP) is equivalent to 100m (2drms) for a ratio of 2.5.

The spherical error probable (SEP) is used to express the accuracy of the three-dimensional case. SEP means that there is a 50 percent chance that the true three-dimensional position is located inside a sphere of a radius equal to the value of SEP [1].

Reference

[1]   Mikhail, E., *Observations and Least Squares,* New York: University Press of America, 1976.

Appendix C:
Useful Web Sites

C.1 GPS/Glonass/Galileo Information and Data

U.S. Coast Guard Navigation Center (GPS NANU, GPS Almanac, FRP, and others):

http://www.navcen.uscg.gov

U.S. Naval Observatory (GPS timing data and information):

http://tycho.usno.navy.mil/gps_datafiles.html

National Geospatial-Intelligence Agency:

www.nga.mil

GLONASS information center:

http://www.glonass-center.ru

Russian Space Science Internet:

http://www.rssi.ru

European Space Agency (Galileo and EGNOS information):

http://www.esa.int/esaCP/index.html

WAAS information:

http://gps.faa.gov/Programs/WAAS/waas.htm

Canadian Wide-Area DGPS service:

http://www.cdgps.com/e/index.htm

Radio Technical Commission for Maritime Services:

http://www.rtcm.org

National Marine Electronics Association:

http://www.nmea.org

GPS World magazine:

http://www.gpsworld.com

U.S. Institute of Navigation (GNSS proceedings):

http://www.ion.org

IEEE publications:

http://ieeexplore.ieee.org

Navtech Seminars and GPS Supply:

http://www.navtechgps.com

IGS (formerly the International GPS Service):

http://igscb.jpl.nasa.gov

International Terrestrial Reference Frame:

http://itrf.ensg.ign.fr

The International Earth Rotation and Reference Systems Service:

http://www.iers.org

Canadian Spatial Reference System of Natural Resources Canada:

http://www.geod.nrcan.gc.ca/index_e.php

Canadian Active Control System Products:

http://www.geod.nrcan.gc.ca/network_a/active_e.php

Precise Point Positioning service:

http://www.geod.nrcan.gc.ca/ppp_e.php

National Transformation (NTv2) Computation:

http://www.geod.nrcan.gc.ca/apps/ntv2/index_e.php

Jet Propulsion Laboratory GPS Satellite Performance:

http://milhouse.jpl.nasa.gov/eng/eng.sat.html

U.S. National Geodetic Survey:

http://www.ngs.noaa.gov

U.S. Continuously Operating Reference Station (CORS data):

http://www.ngs.noaa.gov/CORS

University Navstar Consortium (UNAVCO data and service):

http://www.unavco.org

EUREF Permanent GPS Network:

http://www.epncb.oma.be/euref_IP

C.2 Some GPS Manufacturers

NovAtel (GPS receivers and integrated systems):

http://www.novatel.ca

Thales Navigation (Ashtech precision products):

http://products.thalesnavigation.com/en

Trimble Navigation:

http://www.trimble.com

Leica Geosystems:

http://www.leica-geosystems.com

SOKKIA Corporation:

http://www.sokkia.com

Applanix Corporation (integrated systems):

http://www.applanix.com

Pacific Crest Corporation (radio link systems):

http://www.paccrst.com

About the Author

Dr. Ahmed El-Rabbany obtained his Ph.D. degree in GPS from the Department of Geodesy and Geomatics Engineering, University of New Brunswick, Canada. He is currently working as an associate professor at Ryerson University in Toronto, Canada. He also holds an honorary research associate position at the Department of Geodesy and Geomatics Engineering, University of New Brunswick, and an adjunct professor position at York University. Dr. El-Rabbany has more than 20 years of research, instructional, and industrial experience in the general discipline of geomatics engineering, with specializations in GPS, geodesy, data modeling and estimation, and hydrographic surveying. He leads a number of research activities in the areas of GPS, integrated navigation systems, and hydrographic surveying. Dr. El-Rabbany holds/held a number of professional positions, including Geodesy Councilor to the Canadian Institute of Geomatics; Chair, International Federation of Surveyors (FIG) WG-4.2: Vertical Reference Surface for Hydrography; and chair of the Toronto branch of the Canadian Institute of Geomatics. Dr. El-Rabbany has received a number of awards in recognition of his academic achievements, including two merit awards from Ryerson University. Dr. El-Rabbany is included in the *International Directory of Experts—Satellite Navigation*, issued by the American Biographical Institute.

Index

Accuracy measures, 195–96
Airborne laser bathymetry system, 154
Airborne mapping, 151–52
Almanac, GPS, 39–40
Ambiguity resolution
    antenna swap method, 84, 85–86
    known baseline method, 84
    OTF, 84, 86–88
    parameters, 83
    techniques, 83–88
Angle of arrival (AOA), 134–35
Antennas
    offset, 56
    phase center variation, 49–50
Antenna swap method, 84, 85–86
    defined, 85
    illustrated, 85
    initialization procedure, 85
Applications, 11, 139–61
    airborne mapping, 151–52
    cadastral surveying, 159
    civil engineering, 144–45
    construction, 144
    forestry/natural resource, 141–42
    land seismic surveying, 148–49
    marine seismic surveying, 149–50
    open-pit mining, 146–48
    precision farming, 142–43

    retail industry, 157–59
    seafloor mapping, 152–54
    structural deformation monitoring,
        145–46
    transit systems, 156–57
    utility industry, 139–41
    vehicle navigation, 154–55
    waypoint navigation, 160–61
Applied Physics Laboratory (APL), 1
Arclogistics, 157
Assisted GPS (A-GPS), 135
Autonomous GPS accuracy, 9

Beidou system, 170–72
    defined, 170
    ground segment, 170–71
    operational concept, 172
    as two-way ranging system, 171
BERENSE software, 73
Biases. *See* Errors and biases

Cadastral surveying, 159
Canada-wide differential GPS (CDGPS),
    97
Canadian Active Control System (CACS),
    67
Carrier phase measurements, 22–24
    defined, 23
    illustrated, 23

Cellular
    AOA, 134–35
    communication technology, 134
    GPS integration, 134–36
    limitations, 134
    TDOA, 134–35
Chinese regional satellite navigation system.
        See Beidou system
Civil engineering applications, 144–45
Coarse acquisition (C/A code), 14
Communication (radio) link, 80–81
Continuously Operating Reference Station
        (CORS), 89
Control sites, 7–8
Conventional Terrestrial Reference System
        (CTRS), 179–80
Coordinated Universal Time (UTC),
        20–21
    defined, 20
    GPS time versus, 21
Coordinate systems
    CTRS, 179–80
    defined, 177
    ENU, 179
    geodetic, 177–81
    one-dimensional, 177
    three-dimensional, 178
    two-dimensional, 177
Crustal Dynamics Data Information
        System (CDDIS), 39
Cycle slips, 25–26
    defined, 25
    illustrated, 25
    occurrence, 25
Cyclic redundancy check (CRC), 117

Data and product services, 91–93
Datums, 175–77
    geodetic, 178–79
    transformations, 182–83
    vertical, 177
Dead reckoning (DR)
    defined, 128
    GPS integration, 128–29
    navigation, 128–29
    odometer sensors, 129
    vibration gyroscopes, 129

Differential GPS (DGPS), 46
    beacon service, 93–94
    Canada-wide (CDGPS), 97
    corrections, 115
    global (GDGPS), 97
    maritime service, 93–95
    nationwide (NDGPS), 94–95
    real-time, 77–79
    wide-area (WADGPS), 90, 95–98
Differential positioning. *See* GPS relative
        positioning
Dilution of precision (DOP), 59
Doppler measurements, 24
Doppler shift, 10
Dual-frequency receivers, 19–20

Enhanced Loran (e-Loran), 126
ENU coordinate system, 179
Ephemeris errors, 44–45
Errors and biases, 43–61
    antenna phase center variation, 49–50
    ephemeris, 44–45
    hardware delay, 57
    illustrated, 43
    ionospheric delay, 51–55
    multipath, 48–49
    phase windup, 56–57
    receiver measurement noise, 50–51
    satellite and receiver, 47–48
    satellite antenna offset, 56
    selective availability, 45–47
    site displacement effect, 57
    tropospheric delay, 55–56
European geostationary navigation overlay
        system (EGNOS), 97

Fast (rapid) static surveying, 73–74
    defined, 73
    illustrated, 74
    use, 73–74
    See also GPS positioning
Forestry applications, 141–42

Galileo, 167–70
    commercial service (CS), 167
    constellations, 168
    defined, 167–68
    frequency overlap, 169

ground segment, 169
in-orbit validation element, 169
navigation signal parameters, 170
open service (OS), 167
public regulated service (PRS), 168
safety of life (SOL), 167–68
segments, 168–69
service levels, 167–68
support to search and rescue (SAR), 168
Gas ionization, 51
Geodetic coordinate system, 177–81
Geodetic datum, 178–79
Geographical information system (GIS), 139, 140
Geoid-ellipsoid separation, 192
Geostationary Earth orbit (GEO), 33
GGA, 119–20
   explanation of sentence terms, 120
   sentence structure, 119
Giove-A, 169
Global differential GPS (GDGPS), 97
Global Positioning System. *See* GPS
GLONASS system, 91, 92, 163–67
   autonomous positioning accuracy, 166
   channel numbers, 164–65
   constellation, 165
   constellation status, 166
   control segment, 166
   defined, 163
   GLONASS-K, 166
   GPS integration, 167
   navigation message, 167
   orbital configuration, 164
   receivers, 165
   satellites, 165–66
   segments, 163
GPS
   accuracy measures, 195–96
   almanac, 39–40
   applications, 11, 139–61
   assisted (A-GPS), 135
   autonomous accuracy, 9
   control sites, 7–8
   coordinates obtained with, 181–82
   data and product services, 91–93
   details, 13–27
   ephemeris error, 44–45

errors and biases, 43–61
idea behind, 8–10
integration, 123–36
introduction, 1–11
levels of service, 10–11
linear combinations, 26–27
modernization, 16–17
navigation message, 15–16
observations, 8
overview, 2
performance evaluation, 51, 52
precision measures, 195–96
real-time differential, 77–79
receiver types, 18–20
satellite generations, 4–6
satellite orbits, 2, 29–41
service, 2
signal structure, 13–16
sky plot, 41
standard formats, 103–21
system monitoring, 8
white use?, 11
GPS and GEO augmented navigation (GAGAN) system, 97
GPS broadcast orbit, 36–39
   defined, 36
   parameter explanations, 38–39
   parameters, 37
GPS/cellular integration, 134–36
GPS constellations
   buildup, 4
   current, 6–7
   illustrated, 2
   number of satellites, 3
GPS/dead reckoning integration, 128–29
GPS/INS integration, 130–32
GPS/Loran-C integration, 123–27
GPS/LRF integration, 127–28
GPS manufacturers, 199–200
GPS point positioning, 66–67
   defined, 65, 66
   performance, 67
   principle, 66
   unknown parameters, 66–67
   See also GPS positioning
GPS positioning
   communication link, 80–81

GPS positioning (continued)
    fast (rapid) static surveying, 73–74
    high-accuracy, 89
    idea behind, 9
    kinematic GPS surveying, 76
    modes, 65–81
    multisite RTK, 98–99
    point, 65, 66–67
    PPP, 65, 68–70
    real-time DGPS, 77–79
    relative, 65–66, 70–71
    RTK GPS surveying, 76–77
    static GPS surveying, 71–73
    stop-and-go GPS surveying, 74–76
GPS/pseudolite integration, 132–34
GPS relative positioning, 70–71
    defined, 65–66, 70
    principle, 71
    satellite requirements, 70
    See also GPS positioning
GPS segments, 3–4
    control segment, 3–4
    defined, 3
    illustrated, 34
    space segment, 3
    user segment, 3
Group repetition interval (GRI), 125

Handover word (HOW), 15
Hardware delay, 57
Height systems, 191–92
Horizontal dilution of precision (HDOP),
    59

Ideal satellite orbit, 33–35
Inclined geosynchronous orbit (IGSO), 33
Inertial navigation system (INS)
    defined, 130
    GPS integration, 130–32
    inertial measurement unit (IMU),
        130–31
    MEMS-based technology, 131–32
Integration, 123–36
    GPS/cellular, 134–36
    GPS/dead reckoning, 128–29
    GPS/INS, 130–32
    GPS/Loran-C, 123–27
    GPS/LRF, 127–28

GPS/pseudolite, 132–34
International Association of Geodesy
    (IAG), 91
International Atomic Time (TAI), 20
International GNSS Service (IGS), 45
    GPS data, 92
    tracking stations, 91
International Terrestrial Reference Frame
    (ITRF), 90
Ionospheric delay, 51–55
    defined, 52
    number of free electrons and, 52–53
    removal, 54–55
    TEC and, 53–54
    See also Errors and biases
Issue of data clock (IODC), 38
Issue of data ephemeris (IODE), 38

Japanese satellite navigation system.
    See QZSS system

Kepler
    first law of planetary motion, 30–31
    second law of planetary motion, 31
    third law of planetary motion, 31
Keplerian satellite orbit, 33–35
    defined, 33
    illustrated, 34
    parameters, 34
Kinematic GPS surveying, 76
Klobuchar model, 55
Known baseline method, 84

LAMBDA method, 87
Lambert conical projection, 188–89
    defined, 188
    illustrated, 189
    parallels, 188
    See also Map projections
Land seismic surveying, 148–49
Laser range finders (LRFs), 127–28
Linear combinations, 26–27
    formation, 27
    illustrated, 27
Local arbitrary mapping systems, 190–91
Location-based services (LBS), 135
Loran-C
    autonomous precision, 127

characteristics, 126
defined, 123
GPS integration, 123–27
GPS receiver, 127
individual pulse, 125
operation frequency, 124
principle, 124
pulse pattern, 125
receivers, 125
Low-accuracy positioning, 176
Low Earth orbit (LEO), 33

Mapping
airborne, 151–52
crop field, 143
seafloor, 152–54
utility, 139–41
Map projections, 183–90
concept illustration, 183
defined, 183
Lambert conical, 188–89
modified transverse Mercator (MTM),
187–88
popular, 184
stereographic double, 189–90
Transverse Mercator, 184–85
universal transverse Mercator (UTM),
185–87
Marine seismic surveying, 149–50
GPS for, 149–50
illustrated, 150
principle, 149
results, 150
See also Applications
Maritime DGPS service, 93–95
Master control station (MCS), 3–4, 7–8
Measurement-domain WADGPS, 95–96
Measurements
carrier phase, 22–24
Doppler, 24
errors and biases, 26
pseudorange, 21–22
receiver noise, 50–51
Medium Earth orbit (MEO), 33
Minimum shift keying (MSK), 94
Mission-planning software, 40
Modernization, 16–17

Modified transverse Mercator (MTM)
projection, 187–88
defined, 187
illustrated, 188
scale factor, 187
See also Map projections
Monitoring, structural deformations,
145–46
MTSAT-based satellite augmentation
system (MSAS), 97
Multibeam echo sounders, 154
Multipath
defined, 48
effect illustration, 49
errors, 48–49
Multisite RTK system, 98–99

NAD 83 system, 180–81
National Geodetic Survey (NGS), 93
National Geospatial Intelligence Agency
(NGA), 8
parameters, 182, 183
tracking stations, 8
Nationwide differential GPS (NDGPS),
94–95
Navigation message, 15–16
NMEA 0183 format, 118–21
data streams, 118–19
defined, 118
GGA sentence structure, 119
GGA sentence terms, 120
GNSS-related sentences, 119
See also Standard formats
Notice Advisory to Navstar Users
(NANU), 8

Odometer sensors, 129
OmniSTAR system, 96
On-the-fly. *See* OTF ambiguity resolution
Open-pit mining
applications, 146–48
centimeter level-accuracy guidance, 147
cycle phases, 147
GPS for, 146–48
illustrated, 148
Operational control segment (OCS), 4
Orbits, 2, 29–41
ellipse geometry, 32

Orbits (continued)
 GEO, 33
 GPS broadcast, 36–39
 ideal, 33–35
 IGSO, 33
 LEO, 33
 MEO, 33
 perturbed, 35–36
 semimajor axis, 2
 types, 32–33
OTF ambiguity resolution, 76, 84, 86–88
 confidence region, 86–87
 covariance matrix representation, 86
 defined, 86
 illustrated, 87
 LAMBDA method, 87

Performance
 GPS evaluation, 51, 52
 GPS point positioning, 67
Personal communication services (PCS), 81
Perturbed satellite orbit, 35–36
Phase windup, 56–57
Position Data Link (PDL), 80
Position dilution of precision (PDOP), 59
Position-domain WADGPS, 95
Postprocessing, real-time versus, 79
PPP, 24, 68–70
 defined, 65, 68
 results, 70
 See also GPS positioning
Precise point positioning. See PPP
Precise positioning service (PPS), 10–11
Precision code (P-code), 14–15
Precision farming
 aerial guidance system, 143
 applications, 142
 crop field mapping, 143
 GPS for, 142–43
 soil sample collection, 143
Precision measures, 195–96
Primary phase factor (PF), 126
Pseudolite signals, 132–34
Pseudorandom noise (PRN), 7
Pseudorange correction (PRC), 114
Pseudorange measurements, 21–22
 defined, 22

 illustrated, 22

Quasi-Zenith Satellite System. See QZSS
 system
QZSS system, 172–73
 constellation, 172
 defined, 172
 ITRS as reference frame, 173

Radio Technical Commission for Maritime
 Service (RTCM), 78
Ranging codes, 14
Real time differential GPS, 77–79
Real-time kinematic (RTK), 11
Receivers, 10
 dual-frequency, 19–20
 errors, 48
 features, 19
 first generation, 18
 GLONASS system, 165
 illustrated, 19
 measurement noise, 50–51
 single-frequency code, 19
 tracking channels, 18
 types, 18–20
Repeaters, 81
Retail industry application, 157–59
 components, 157
 real-time fleet-monitoring system,
  158–59
 routing timing, 158
RINEX format, 36, 39, 91
 data section, 106
 defined, 104
 files, 105–6
 meteorological file, 107, 108
 navigation file, 107, 108
 observation file, 105–6
 versions, 104–5
 See also Standard formats
RTCM SC-104 standards, 112–18
 backward compatibility, 118
 defined, 112
 DGPS corrections, 115
 first-word decoding example, 116
 layers, 118
 message information, 114
 message readability, 114

message structure, 115, 116
message types, 113–14
raw corrections, 116
versions, 112
weaknesses, 117
See also Standard formats
RTK GPS surveying, 76–77
defined, 76
illustrated, 77
on-the-fly (OTF) ambiguity resolution, 76
positioning accuracy, 77

Satellite-based augmentation systems (SBAS), 97
Satellites, 8–9
antenna offset, 56
attitude error, 56
Block IIF, 5–6
Block III, 5–6
Block IIR, 5
errors, 47–48
first generation, 4
generations, 4–6
geometry measures, 57–60
GLONASS system, 165–66
orbits, 2, 29–41
ranging codes, 14
second generation, 4–5
transmission, 9
visibility, 40–41
Seafloor mapping, 152–54
airborne laser bathymetry system, 154
GPS for, 152–54
illustrated, 153
multibeam echo sounders, 154
single-beam echo sounder, 153
water depth, 153
See also Applications
Secondary phase factor (SF), 126
Selective availability, 45–47
defined, 45
errors, 45–46
Short baseline test, 52
Signal structure, 13–16
Single-frequency code receiver, 19
Site displacement effect, 57

Sky plot, 41
SP3 format, 109–12
data section, 110, 111
defined, 109
header section, 110
position and clock correction record, 112
precise ephemeris file, 109–10
standard deviation computation, 111
See also Standard formats
Space object motion, 29–32
Space vehicle number (SVN), 6
Stakeout. *See* Waypoint navigation
Standard formats, 103–21
NMEA 0183 format, 118–21
RINEX format, 104–9
RTCM SC-104, 112–18
SP3 format, 109–12
Standard positioning service (SPS), 10
State-space-domain WADGPS, 96–97
Static GPS surveying, 71–73
with carrier-phase measurements, 73
defined, 71
illustrated, 72
Stereographic double projection, 189–90
cases, 189–90
defined, 189
illustrated, 190
perspective point, 190
See also Map projections
Stop-and-go GPS surveying, 74–76
defined, 74–75
illustrated, 75
initial integer ambiguity parameters, 75
kinematic, 76
receivers, 74–75
Structural deformation monitoring, 145–46
Surveying
cadastral, 159
fast (rapid) static, 73–74
kinematic, 76
land seismic, 148–49
marine seismic, 149–50
RTK GPS, 76–77
static, 71–73
stop-and-go, 74–76

Time-difference of arrival (TDOA), 134–35
Time difference (TD), 125
Time dilution of precision (TDOP), 59
Time systems, 20–21
Time to first fix (TTFF), 135
Total electron content (TEC), 53–54
Transit systems application, 156–57
Transverse Mercator projection, 184–85
    defined, 184
    illustrated, 185
    imaginary cylinder, 184–85
    secant cylinder, 185
    tangent cylinder, 185
    See also Map projections
Triple difference, 25, 26
Tropospheric delay, 55–56
    defined, 55
    dependencies, 55
    dry/wet, 55–56
    See also Errors and biases

Universal transverse Mercator (UTM)
        projection, 185–87
    defined, 185
    illustrated, 186
    negative coordinates and, 187
    scale factor, 186
    zoning, 186
    See also Map projections
User equivalent range error (UERE), 60–61
Utility mapping, 139–41

GIS for, 140
GPS for, 141
See also Applications

Vehicle navigation application, 154–55
Vertical datum, 177
Vertical dilution of precision (VDOP), 59, 125
Vibration gyroscopes, 129

Waypoint navigation, 160–61
    defined, 160
    guidance information, 161
    illustrated, 160
Web sites, 197–200
    GPS/Glonass/Galileo information/data, 197–99
    GPS manufacturers, 199–200
WGS 84 system, 67, 180, 181
Wide area augmentation system (WAAS), 97
Wide-area differential GPS (WADGPS), 90, 95–98
    categories, 95
    defined, 95
    measurement-domain, 95–96
    position-domain, 95
    principle, 96
    state-space-domain, 96–97

Zero baseline test, 51